Leveling Up as a Tech Lead

Growing as a Technical, Project, and People Leader

Anemari Fiser
Foreword by Patrick Kua

O'REILLY®

Leveling Up as a Tech Lead

by Anemari Fiser

Published by O'Reilly Media, Inc., 141 Stony Circle, Suite 195, Santa Rosa, CA 95401.

O'Reilly books may be purchased for educational, business, or sales promotional use. Online editions are also available for most titles (*http://oreilly.com*). For more information, contact our corporate/institutional sales department: 800-998-9938 or *corporate@oreilly.com*.

Acquisitions Editor: David Michelson	**Indexer:** Sue Klefstad
Development Editor: Shira Evans	**Cover Designer:** Susan Thompson
Production Editor: Gregory Hyman	**Cover Illustrator:** Susan Thompson
Copyeditor: Liz Wheeler	**Interior Designer:** Monica Kamsvaag
Proofreader: Piper Content Partners	**Interior Illustrator:** Kate Dullea

December 2025:   First Edition

Revision History for the First Edition

2025-11-19:   First Release

See *http://oreilly.com/catalog/errata.csp?isbn=9781098177515* for release details.

978-1-098-17751-5

[LSI]

Contents

Foreword

In my first role as a tech lead, a long time ago now, it felt like I was given a blank map and told to chart a course through unknown lands. I had no guide, no proven path, and no certainty that I was heading in the right direction. I had no idea what "good" looked like in this role. Somehow, I must have done something right, because a few years later, I was asked to run tech lead training for my colleagues at Thoughtworks, which eventually grew into a global program. During that time, I realized my struggles were not unique. Many others faced the same uncertainty with very few resources to support them. When I searched for books to recommend, I found only three. Two were management-focused books, and only one of those, *Becoming a Technical Leader* by the late Gerald Weinberg (Dorset House), truly addressed technical leadership. More than a decade ago, I wrote *Talking with Tech Leads* (CreateSpace) to help fill that gap.

Since publishing that book, I have trained thousands of tech leads, engineering managers, and staff engineers and coached dozens of CTOs. Over that time, our industry has shifted dramatically. We moved from on-premises to cloud-based software, from desktop to web to mobile, and from big-bang releases to continuous delivery. Yet, despite these changes, newly appointed tech leads keep asking the same questions, such as *What is expected of me? Where should I be spending my time? How much hands-on coding should I be doing?*

While these fundamental questions remain constant, the role itself has also become far more demanding. Some tech leads guide teams with a shared technical background, such as a team of Java developers. Others lead cross-functional teams that demand knowledge across fields like web, mobile, backend, data science, and more. Even a so-called "full-stack developer" cannot be an expert across such a broad range, and the rapid adoption of AI-driven tools has only increased the complexity. The confusion surrounding the tech lead role has deepened with the emergence of newer roles such as engineering manager and staff engineer, where responsibilities often overlap. On top of all this, many teams now follow a "You build it, you run it" model, adding operational and support work to the tech lead's responsibilities. The modern tech lead role is more multifaceted than ever before.

Despite the increased complexity, dedicated resources for tech leads remain rare. Our industry has produced excellent books like *The Manager's Path* (Camille Fournier), *The Staff Engineer's Path* (Tanya Reilly), and *The Engineering Executive's Primer* (Will Larson) (all O'Reilly). But tech leads still lack the same breadth of practical guidance, which is why I am excited about this book.

When Anemari asked me to write this foreword, she reminded me that she took part in one of my tech lead courses in Barcelona, which helped her grow into the role. This book is a testament to her continued growth, and in its pages, she shares the lessons, experiences, and tools that helped her succeed.

She begins with the most pressing and complicated question: *What is the scope of the tech lead role?* It is a difficult question because every organization sets different expectations. Even when role descriptions exist, each team and situation demands a different approach, something described as situational leadership. Anemari provides a starting point to help you define what matters most in your current situation.

From there, the book lays out a map of common responsibilities across both people and technical domains. On the people side, you will learn how to build relationships, foster a strong feedback culture, and guide your team toward high performance. On the technical side, you will explore how to understand, define, and improve your system architecture, align technical decisions with business goals, and cultivate strong technical practices that allow your team to make continuous changes to your systems with confidence.

In addition to these responsibilities, Anemari highlights common challenges every tech lead will face. These include avoiding micromanagement, navigating difficult conversations, and addressing technical debt. For each, she offers practical strategies you can apply immediately, grounded in her own experience.

With the ideas, tools, and experiences Anemari shares, I am confident this book will help you navigate the uncertainty of the tech lead role and prepare you to lead with intent. No book can give you every answer, but this one offers you the tools to chart your own path as a tech lead. And that, in the end, is what leadership is all about.

—Patrick Kua, CTO coach,
founder of the Tech Lead Academy,
coauthor of Building Evolutionary
Architectures *(with Neal Ford,*
Rebecca Parsons, and Pramod
Sadalage; O'Reilly), and author
of Talking with Tech Leads
and The Retrospective Handbook
(both CreateSpace)
Berlin, Germany (August 2025)

Preface

Early in my career, I realized I wanted to become a tech lead, although at the time, my understanding of the role was limited. What motivated me most was the impact we could have as a team rather than my individual contributions.

While the engineers around me were diving deep into the latest technologies and tackling more complex technical challenges, I found myself questioning the value of the work we were doing. I was more drawn to solving team-related problems and taking the initiative to move things forward. I wanted to have more influence on how the team operated, so when the scrum master left, I saw an opportunity and stepped into that role. While others were focused on building new features, I constantly found myself documenting the existing ones, improving communication, and streamlining our processes.

To make it happen, I moved to another country and joined Thoughtworks: a company that supported my growth into the tech lead role. I started working with my tech lead at the time to develop the necessary skills, and I said yes to every training opportunity that came my way. When I learned that the team's tech lead was leaving, I asked to take over. They said yes.

Once I realized that the type of impact I wanted could happen only in the tech lead role, I began focusing on the skills I believed were necessary for the position. I worked hard to sharpen my technical abilities, thinking that to lead a technical team, I needed to be the most technical person on the team. I was wrong.

The more I learned about the tech lead role, the more I felt like I didn't know. There are just so many opinions out there on what the role actually is. The more I talked to other tech leads about it, the more I realized they didn't know either. Everyone seems to have their own definition, but the tech industry cannot agree on what exactly the role is.

This confusion isn't unique to the tech lead role; it applies to most roles in tech, from junior developer to CTO, or anything in between. But when it comes to the tech lead role, things get even more fuzzy. As proof, I couldn't find a single book out there focusing solely on it. There are plenty of general engineering leadership books, but as a tech lead, it's up to you to figure out what's relevant to your specific situation. This can get overwhelming quickly, and I definitely felt that struggle when I was starting out.

Why I Wrote This Book

I wrote this book to save other tech leads, like you, from the painful process of going through countless resources, trying to figure out what applies to your specific role. You can think of it as your go-to guide for navigating the unique challenges of being a tech lead.

This book draws from my personal experience as a tech lead at Thoughtworks, insights from other more and less seasoned tech leads I've worked with, and what I learned from training and coaching over 300 tech leads across various companies, cultures, and environments over the past three years.

How to Effectively Use This Book

No matter where you are in your journey as a tech lead—just starting out, mid-career, or very experienced—this book provides valuable insights, tools, and lessons. If you're an aspiring tech lead, aiming to step into the role, I recommend starting with the first two chapters:

Chapter 1, "The Role of a Tech Lead"
This chapter gives you a clear, high-level picture of what the role actually involves. You'll explore the day-to-day responsibilities, how to define the scope of the role in your context, what's expected of you, and the different paths that can lead to becoming a tech lead. This chapter also helps you assess if the role is a good fit for you right now.

Chapter 2, "How to Become a Tech Lead"
Once you've decided you're ready (or almost ready) to make the move, Chapter 2 helps you build a personal growth plan. It walks you through mindset shifts, how to assess your current strengths and gaps, and what steps to take to grow into the role with intention. It also highlights the common challenges new tech leads face and how to navigate them early on.

It's also worth scanning through the entire book to familiarize yourself with the common scenarios you'll likely encounter in the role.

If you've just stepped into the role, this book will be your go-to reference for troubleshooting and solving the daily challenges you'll face. It's packed with real-world scenarios, so you can easily find the relevant chapters and get practical advice and ideas for handling specific issues.

If you've been in the role for a while now, this book can help you refine your leadership skills by identifying areas for improvement, and discover new strategies for dealing with the ongoing challenges of the role.

The more I was learning about the role from training, observing my experienced tech lead, and talking with others in the role, the clearer it became: being a tech lead is less about tech and more about people. When I finally stepped into the role myself, it all clicked: the tech lead is 100% more about people than tech.

Reaching this conclusion was a long and challenging journey of self-development and learning, a process I see tech leads go through every day. That's why this book addresses both the technical aspects and the equally important people and business side of the role.

Conventions Used in This Book

The following typographical conventions are used in this book:

Italic

> Indicates new terms, URLs, email addresses, filenames, and file extensions.

O'Reilly Online Learning

O'REILLY® For more than 40 years, *O'Reilly Media* has provided technology and business training, knowledge, and insight to help companies succeed.

Our unique network of experts and innovators share their knowledge and expertise through books, articles, and our online learning platform. O'Reilly's online learning platform gives you on-demand access to live training courses, in-depth learning paths, interactive coding environments, and a vast collection of text and video from O'Reilly and 200+ other publishers. For more information, visit *https://oreilly.com*.

How to Contact Us

Please address comments and questions concerning this book to the publisher:

O'Reilly Media, Inc.

141 Stony Circle, Suite 195

Santa Rosa, CA 95401

800-889-8969 (in the United States or Canada)

707-827-7019 (international or local)

707-829-0104 (fax)

support@oreilly.com

https://oreilly.com/about/contact.html

We have a web page for this book, where we list errata, examples, and any additional information. You can access this page at *https://oreil.ly/leveling-up-tech-lead*.

For news and information about our books and courses, visit *https://oreilly.com*.

Find us on LinkedIn: *https://linkedin.com/company/oreilly-media*.

Watch us on YouTube: *https://youtube.com/oreillymedia*.

Acknowledgments

First, I want to thank my partner, Radu Chilom. You've been there the whole time, through deadlines, meltdowns, and ups and downs. You encouraged me, debated leadership scenarios and technical stories with me at midnight, and kept me sane when the process made me a little crazy. Thank you!

To my friends, Elena Garcia, Stefania Rosca, Sorin Mihai, Eduard Almasque, Elisa Cutrin, Valentina Servile, thank you not just for your thoughtful feedback on the book but for keeping my spirits high with advice, encouragement, and friendship. Many of you I met at Thoughtworks, and I'm grateful for every moment there. The people, the international network, the opportunities, and the trust pushed me out of my comfort zone and shaped the way I lead today. Much of what's in this book is rooted in those experiences.

A huge thank-you to everyone who showed interest in reviewing the book. When I asked for volunteers, I definitely didn't expect more than 160 people, including friends, former colleagues, and even people I only knew from

LinkedIn. I couldn't take everyone into the official reviewing process, but every application and comment boosted my confidence that this book was needed.

Which brings me to my tech reviewers. Special thanks to those who read carefully, shared their stories and experience, debated topics with me, and made the book so much better: Alex Geogea, Sonu Kapoor, Jeff Zinger, Hermann ("Ham") Vocke, Kaitlyn Tierney, Andra Popa, Ivo Pinto, Sergio Visinoni, Joe Seymour, Mireia Angles, Michael Di Prisco, Mihaela Pasculescu, Aleix Morgadas, Yanqi Luo, Alex Lau, Rodrigo Borrego, Dani Roman, Nicolas Gonzalez Avalis, Dariusz Sadowski, Dagna Bieda.

To all the companies who trusted me to grow their tech leads, and to the tech leads themselves, whose challenges, questions, and stories inspired so much of this book, thank you. Every conversation shaped these pages.

A warm thank-you to Pat Kua for your generous foreword and endorsement of this book, and to Birgitta Böckeler for your insights and collaboration on the AI sections, your thoughtful contributions made it stronger and more complete.

And last but not least, my gratitude to the team at O'Reilly. It was a pleasure working with each and every one of you. David Michelson, who got me on board, my editors Jill Leonard, Gregory Hyman, and especially Shira Evans, thank you for your guidance, accountability, and support. You took my messy drafts and helped me shape them into something I'm truly proud of. And to all the other people who contributed in ways I may never know about, thank you.

The Role of a Tech Lead

The tech lead role is one of the most misunderstood positions in the tech industry. It's a role that exists at the intersection of people, technology, and business, and yet there's little consensus on what it actually means. This lack of clarity often leaves new tech leads feeling unsure about what's expected of them and how to approach the role effectively.

Every company seems to have its own version of the tech lead role, shaped by their culture, processes, and priorities. Some see the role as deeply technical, while others focus heavily on team dynamics and processes. Understanding these differences and defining what it means within your specific context is the first step to succeeding as a tech lead.

That's why, in this chapter, I'll outline the essential aspects of the tech lead role to help you understand what to expect and how to approach it. I'll begin by defining what a tech lead is, including the scope and daily responsibilities. Then, I'll explore how to make sense of the expectations placed on you, whether you have a formal job description or not, and offer guidance on evaluating whether this role is right for you.

I'll also look at the different paths people take to become a tech lead, from transitioning within your current team to forming entirely new ones. Finally, I'll cover the critical skills and mindset shifts you'll need to go through in this transition.

What Is a Tech Lead?

When I ask tech leads, "What's expected of you?" the responses are wildly different, covering a wide range of skills and responsibilities.

At one extreme, some describe the tech lead as the most technical person on the team, someone with deep expertise in the tech stack who makes all the decisions and contributes code daily, often at the same level as any other engineer. The role is seen as highly hands-on, deeply technical, and rooted in direct contributions to the team's output.

At the other end of the spectrum, the tech lead is described as someone who rarely codes, either due to lack of time or because they deliberately choose to focus on other priorities, like people, processes, and shielding the team from distractions. In this view, the tech lead is more of a facilitator, working closely with the product manager (PM) to manage the backlog and ensure smooth collaboration.

Figure 1-1 shows a word cloud compiled from typical responses I get when training groups of tech leads and asking, "What words do you associate with 'tech lead'?"

Figure 1-1. Common words associated with "tech lead"

These examples show just how varied the expectations can be, from being a hands-on technical expert to focusing entirely on people and processes. The reality is that the tech lead role sits somewhere in the middle, requiring a careful balance between technical leadership, team development, and stakeholder alignment. That balance isn't fixed. It can shift over time, depending on the maturity of your team, the phase of the project, or even the specific challenges you're facing.

A tech lead is expected to guide the team through decision making, ensuring that everyone's input is heard and considered, rather than just making decisions unilaterally. This requires a breadth of knowledge, not just about technical areas like infrastructure, architecture, and the tech stack but also about the business context and stakeholder management.

At the same time, a tech lead must support the team's delivery while helping individual team members grow. It's a role that demands both technical expertise and people skills, combining them effectively to build alignment, encourage teamwork, and deliver impactful results.

This wide range of interpretations of the role can be confusing, especially for new tech leads, but it also highlights why the role is so critical, and why defining it within your context is essential.

DEFINING THE SCOPE

Given how widely the tech lead role can vary, it's helpful to look at how different experts and industry leaders define it. Each brings a slightly different lens, technical, organizational, or people-focused, reflecting the fluid nature of the role across different companies and team setups.

Camille Fournier, in *The Manager's Path* (O'Reilly), describes the tech lead as a senior individual contributor who helps a team of engineers work together effectively. She emphasizes coordination, communication, and supporting the team in delivering high-quality work. As she explains it, the tech lead isn't necessarily the strongest coder but the one ensuring that things move forward smoothly.

Will Larson, in *Staff Engineer* (*https://staffeng.com/book*) (self-published), presents the tech lead as an archetype of staff-plus roles: "The Tech Lead guides the approach and execution of a particular team. They partner closely with a single manager, but sometimes they partner with two or three managers within a focused area. Some companies also have a Tech Lead Manager role, which is similar to the Tech Lead archetype but exists on the engineering manager ladder and includes people management responsibilities." His framing emphasizes the tech lead's proximity to execution and operational leadership.

Pat Kua, in his book *Talking with Tech Leads* (CreateSpace), defines the role as "a leader, responsible for a (software) development team, who spends at least 30 percent of their time writing code with the team." His definition underscores the hybrid nature of the role: still technical but with an added layer of responsibility for team effectiveness and communication.

Pat's work had a lasting influence on how I approached the role early in my career. One of my first tech lead trainings was with him, back when we were both at Thoughtworks, and many of the lessons from that experience continue to shape how I teach and support tech leads today.

These perspectives highlight different, but overlapping, facets of the tech lead role: from deep technical execution to team coordination and stakeholder alignment. And that diversity reflects what I've consistently observed in practice.

Drawing from my own experience as a tech lead, and from training hundreds of tech leads over the years, I've seen this role take shape in many forms. The tech leads I've worked with have come from a wide range of environments: product companies and consultancies, startups and large enterprises, and teams at varying levels of maturity. This includes fintech, ecommerce, and SaaS (software as a service) organizations; agile and less agile teams; both layered and flat hierarchies; and setups ranging from internal tooling to customer-facing product teams.

Based on this broad exposure and firsthand experience, I've come to see a clear throughline in what great tech leads do. While the role flexes based on context, there's a common core. Here's how I define it:

The tech lead is a software engineer responsible for leading a development team and accountable for the technical deliverables of that team.

Being accountable for technical deliverables also means ensuring that the team's work aligns with stakeholder expectations. This alignment places the tech lead role right at the intersection of people, business, and technology. It's a balancing act that requires understanding both the technical and nontechnical aspects of the role, as shown in Figure 1-2.

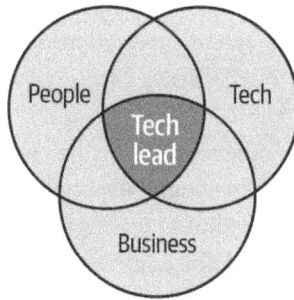

Figure 1-2. What a tech lead is

So, to be an effective tech lead, you need to successfully implement several tasks, described in the following sections.

Build a strong, high-performing team

First, you need to focus on building a strong, high-performing team. It took me a while to figure out what that really means, but here's what I've learned: a strong team delivers value consistently. They work together like a well-oiled machine, complementing each other's skills and continuously growing. They don't wait to be told what to do; they take initiative, and responsibility is shared across the team.

To build and maintain a team like this, you'll need to invest time and energy into your people. That means mentoring and coaching them, creating opportunities for growth through delegation, and providing feedback that actually helps them improve. It also means supporting collaboration by addressing conflict, facilitating open conversations, and building strong relationships. It's challenging work, but it's worth it. Because without effective collaboration, you don't have a team; you just have a group of individuals trying to work together. More on this in the section "Building a High-Performing Team" on page 180.

But there's another layer here that's often overlooked: as a tech lead, you'll also play a critical role in shaping career growth and performance outcomes for your team. While you might not be the one making final decisions on things like promotions or compensation, your input carries weight, especially during performance review cycles.

Your role in performance reviews and career development can vary widely depending on your company's structure and culture. In some organizations, tech leads are deeply involved in performance review cycles, mentoring team members toward the next level, providing structured feedback, and partnering closely with engineering managers to support individual growth. In others, tech leads are barely involved at all, as the expectation is that team leads or engineering managers will take care of the heavy lifting when it comes to reviews and development.

Still, even in companies where your formal involvement is limited, your perspective and feedback are often sought, especially when it comes to evaluating technical growth, impact, and readiness for more responsibility. In many cases, your input may also be requested for engineers outside your team, particularly when you've collaborated on cross-functional work. Promotion cycles can be intense, and tech leads often find themselves at the center of them, writing feedback, assessing performance, and advocating for the people they work with every day.

For a deeper dive into this part of the role, including how to navigate performance reviews, see the section "How to Approach Performance Reviews" on page 197.

Lead technical topics

As a tech lead, you need to ensure your team can deliver on expectations and handle any technical challenges thrown their way. But this doesn't mean you need to be the most technical person on the team or take on every technical decision, code review, or pull request yourself.

A strong technical background is certainly important, and some familiarity with the technologies your team uses is definitely helpful. However, you don't need to have deep expertise in every part of the tech stack. What's important is having enough knowledge to code alongside your team when necessary, staying informed about code changes, and effectively troubleshooting issues that arise. This foundation allows you to guide your team in building a technical strategy, making informed decisions, and solving production problems.

The good news is, you don't need to carry the weight of technical expertise on your own. As a tech lead, you have tools at your disposal to share responsibility and create a balanced workflow. Delegation is key. Delegating tasks to team members who have deeper expertise in certain areas, or who are eager to learn, ensures the workload is distributed effectively while supporting their growth. This approach not only increases your team's overall technical capabilities but also reinforces collaboration and trust.

Equally important is ensuring commitment to the technical strategy and the results you aim to achieve. A great way to do this is by facilitating open, productive conversations within your team. Ensure that everyone has a chance to contribute, share their ideas, and be involved in decision making.

I've seen this approach work firsthand. I once led a team of six developers to deliver a major feature with high business impact on time, under a tight three-month deadline, despite having limited knowledge of the tech stack and business context, no prior relationships within the company, and multiple dependencies on other teams. Instead of trying to become the technical expert, I leaned on team members with deeper expertise in the technology. My focus was on unblocking the team, addressing pain points like backlog alignment, consistent refinement, knowledge sharing, and creating a process that enabled us to deliver value continuously. By fully utilizing everyone's skills and providing clarity on what to work on and how to approach tasks, we achieved our goals together.

Leading technical topics isn't about knowing everything or doing everything; it's about enabling your team to deliver efficiently on expectations.

Bridge tech and business

I'd argue this is one of the most important responsibilities you have as a tech lead because it's something only you can do. You're the go-to person for your stakeholders, and while other developers on your team can take the lead on technical strategy or mentoring, bridging tech and business isn't something you can delegate. It's your job to make your team's successes visible and proactively communicate potential blockers, risks, or delivery impacts to stakeholders. That includes translating technical decisions, like technical debt, refactoring needs, or scaling constraints, into terms that make sense to nontechnical partners.

To do this well, you need to constantly invest in building relationships with your team's stakeholders, rather than just waiting for them to come to you. Set up regular one-on-ones, progress updates, or tracking processes to provide visibility into your team's work. I have seen too many tech leads suffering because they overlooked this, forgetting how much these relationships can influence your team's direction and your ability to lead effectively.

Your role doesn't stop there. You're also responsible for bringing back any relevant information to your team, whether it's feedback, shifting priorities, or risks. And when needed, you have to shield your team from external pressures that might disrupt their focus or morale. This balance between keeping stakeholders informed and protecting your team ensures everyone stays aligned and focused. It's a fine line, but when done right, it can make a huge difference in how your team performs and how smoothly everything runs.

By balancing these three core responsibilities, building a strong team, leading technical topics, and bridging tech with business, you're setting your team up for success. Each of these areas demands attention and focus, as they are all interconnected. A high-performing team requires a clear technical strategy and alignment with business goals. There is no way around it.

TECH LEAD VERSUS TEAM LEAD VERSUS LEAD DEVELOPER VERSUS ENGINEERING MANAGER

Throughout the writing of this book, I received a lot of feedback questioning whether I was actually describing a team lead, a lead developer, or even an engineering manager (EM). The confusion is completely justified.

Different companies define these roles differently. Titles are often inconsistent, and responsibilities are distributed in ways that reflect the company's size, culture, and maturity. But after playing this role with different clients, as well as training and coaching over 300 tech leads from companies of all shapes and sizes over the past three years, I keep coming to the same conclusion:

The tech lead role is a people management role.

Even if the job description doesn't say so. Even if it doesn't sound like it. Even if some techies don't like it.

Because while *tech* is the first word in the title, it's the word *lead* that truly defines the role.

Anyone who has stepped into the tech lead position quickly realizes that the job involves far more people-related work than they expected. I've heard many say:

"This role is way more people-focused than I thought."

"This is not what I signed up for."

"I don't want to be responsible for people's growth."

The reality is that even if a tech lead is officially responsible only for delivery, product quality, and technical decision making, these responsibilities are impossible to fulfill without managing people effectively. How do you ensure quality and delivery without regular check-ins? How do you grow technical excellence without giving feedback, resolving conflict, and motivating others?

So, let's clear up some common misunderstandings that can arise from the various job titles and roles that overlap with being a tech lead.

Tech lead

The tech lead is a software engineer who leads a technical team and is accountable for the outcomes of that team. The keyword is *accountable*. When something breaks or slips, the tech lead is the one who answers for it. That's why they're involved in everything that affects those outcomes: planning, delivery, technical guidance, mentoring, stakeholder management, and people growth.

Even if your company doesn't explicitly define this as people management, it is people management in practice.

Tech lead versus team lead

The team lead role is often conflated with the tech lead. Traditionally, people think of the tech lead as handling technical direction and the team lead as focusing on people. But in most teams I've seen, there isn't a strict split; it's the same person doing both jobs under one title called one or the other.

Some companies might formally separate them, but it's rare to see both roles within a single team. In practice, the difference is mostly semantic. Even if both roles do exist in the same team, it very often happens that they struggle to align due to significant overlap. To avoid stepping on each other's toes, they need to establish very clear boundaries, and that takes strong communication and alignment skills on both sides.

Tech lead versus lead developer

The lead developer is a progression from senior developer. The focus is still highly technical, but the impact is broader, usually going beyond the current team. Lead developers drive initiatives, contribute to architecture, and mentor others, often across multiple teams. They might take on responsibilities like chapter lead or security champion, extending their influence without formally leading a team.

In terms of hierarchy, lead developers are often at the same level as tech leads, just with a different scope: where the tech lead is accountable for a single team's output, the lead developer often works horizontally and is responsible for leading different initiatives, working closely with the tech lead and supporting them in raising the overall technical quality.

Tech lead versus engineering manager

The engineering manager (EM) role is another title with highly inconsistent definitions across companies, and it often overlaps with or absorbs responsibilities commonly associated with tech leads. While both roles involve leadership, the nature of that leadership is different.

Tech leads are deeply embedded in the team, driving technical delivery, mentoring engineers, and ensuring the day-to-day runs smoothly. EMs, on the other hand, tend to operate at a broader level. They often support multiple teams, focus on people development, manage compensation, and shape long-term team health and organizational strategy. In setups where each team has a tech lead, this division works particularly well: the EM supports the tech leads, while the tech leads support the engineers.

In some organizations, tech leads take on all leadership responsibilities because there are no engineering managers. In others, engineering managers handle both people and technical leadership because there are no designated tech leads. The setup varies greatly, and there's no one-size-fits-all model, only what works best for the team and the context.

I see the EM role as a natural career progression in tech after the tech lead role. You can read more about the EM role and how it compares to the tech lead path in the section "Advancing to Engineering Manager" on page 344.

DAY-TO-DAY RESPONSIBILITIES

Now that you have a better understanding of a tech lead's responsibilities, let's translate them into what a typical day for a tech lead might look like.

Meetings

Yes! You'll be spending a lot of time in meetings. There's no way around it. No matter how good your async communication skills are, meetings are essential for alignment (standups, plannings, retros, strategy sessions, discoveries), supporting your team (one-on-ones, feedback sessions, addressing conflicts), and staying informed about changes (all-hands meetings).

But that doesn't mean meetings should consume all your time. Over time, you'll learn which ones truly require your presence, which can be delegated to others, and which can be cut altogether. This allows you to balance your time with two other equally important areas: coding and thinking time.

Coding

I made the mistake once of staying away from the code for too long, and one day, I jumped back in only to realize I didn't even recognize it anymore. Trust me, you don't want to get to that point.

It's easy as a tech lead to get swept up in meetings and high-level discussions, but staying hands-on, coding alongside your team weekly, makes all the difference. It keeps you connected to the work, helps you stay in tune with code quality, and ensures you're better equipped to make technical decisions together with the team.

Pairing with team members is another great way to stay close to both the code and the people. It helps you spot issues early and understand how each developer is progressing, and serves as a powerful mentoring tool. Beyond pairing, reviewing others' code, looking through the codebase, and regularly checking pull requests are also useful ways to stay updated, even if you're not writing much code yourself.

And don't forget about architecture and system design. These are key areas where tech leads are expected to provide guidance and oversight. Staying involved in design discussions keeps your technical context sharp and ensures the team is making aligned, scalable decisions.

Another key area that takes up a surprising amount of time is writing and reviewing technical documentation and plans. Whether you're drafting a design doc, reviewing an RFC (request for comments), or commenting on a team member proposal, this kind of written work is essential to maintain alignment and drive thoughtful technical decision making. It keeps you connected to the work, helps you stay in tune with code quality, and ensures you're better equipped to make technical decisions together with the team.

One important thing to mention here is that, as your coding time naturally decreases and your focus becomes more dynamic, you should avoid becoming a bottleneck for your team. Try not to take on critical tasks that have many dependencies or are on the critical path.

Thinking time

This is something many tech leads tend to overlook: stepping away from the team and giving yourself some quiet time to reflect. Taking this time to think lets you step back, look at how things are going, spot gaps, and plan the team's strategy moving forward. It's one of those things that can save you from being blindsided when things go off track.

For me, Friday afternoons became my go-to time for this. The week was winding down, the chaos was usually settling, and I'd use the time to review my notes, reflect on the week, and figure out what I might be missing. It was also the perfect moment to plan for the next week. This practice was a game-changer:

it gave me clarity, helped me feel more in control, and stopped me from just getting swept up in the day-to-day flow.

Of course, some days will be different than others. Some tech leads prefer saving days just for coding, while others might spend a full day planning. And then there are days when everything goes off track, and you're just putting out fires. What's important is that you're aware of these various activities and ensure each gets the proper attention throughout the week.

In conclusion, if you're stepping into the role for the first time, expect to navigate ambiguity, resolve conflict, influence without authority, and spend as much time on communication as you do on code.

Job titles may differ and responsibilities may be shuffled around, but the bottom line is:

> *No matter how technical the tech lead role may look on paper, it is fundamentally about leading people.*

I'll be honest. I still find it hard to fully embrace this truth every single day. There are moments when I question myself, wondering if I've overstated the people part. But then I talk to another tech lead, about their daily challenges, their struggles, their people issues, and my conviction comes back again: unfortunately, I am right.

Even if some people don't want to hear it. Even if it's hard to accept.

And maybe that's part of the problem. If we were more up-front about just how much people leadership is required to succeed in this role, there would be far less confusion, and far fewer tech leads left wondering why the job feels so hard.

Because most of the struggle comes from underestimating this part of the work.

Understanding the Expectations of Your Role

With so many assumptions about what the tech lead role involves, ranging from being the most technical person on the team to focusing entirely on people and processes, it's impossible to meet every expectation. Some view the role as highly hands-on, expecting the tech lead to contribute code daily and make all the decisions. Others view it as a facilitator's role, focused on guiding processes, supporting the team, and encouraging collaboration.

The key to being effective is understanding your specific context, including your company, team, and stakeholders, and what they expect from you. This clarity will help you define your priorities as a tech lead and focus on what truly matters in your environment.

READ THE JOB DESCRIPTION

Most tech leads I talk to have never read the official company job description for their role. They just made their own assumptions about what is expected of them. No surprise they often ended up in conversations with their managers on how they "are not focusing on the right thing."

So, your first task when jumping into the tech lead role is to read the job description and understand what the people around you expect from you.

REFLECT ON WHAT APPLIES TO YOU AND HOW

Job descriptions for tech lead roles are often vague, filled with phrases like *adaptability and openness to new ideas, grow team members,* or *come up with innovative solutions.* While these may sound inspiring, they leave a lot open to interpretation. To make these expectations meaningful, take the time to reflect on how they apply to your day-to-day work and the current stage of your team.

Your gut feeling is not enough to assess whether you're doing a good job as a tech lead. To truly understand your role and align with others, you need clarity and agreement from your team and stakeholders.

Start by asking the right questions. In one-on-ones with team members, ask, "What do you expect from me as a tech lead?" Juniors might say they want mentoring, guidance, and technical expertise. Seniors, on the other hand, may expect alignment, support, and help in removing blockers. Both perspectives are important, so note their inputs and adapt your approach.

Don't stop with your team. Extend the conversation to other stakeholders like clients, product managers, or department leads. Ask them the same question and listen closely to their answers. They may highlight areas you hadn't considered, like managing external dependencies, aligning with broader business goals, or improving communication across teams.

You can also make this process simpler by using a form to collect responses from multiple people at once. Adding prompts like "What's one thing I can do to better support you?" or "What areas do you think I should focus on as a tech lead?" can help you get more specific and useful feedback.

Once you've gathered input, it's time to synthesize these expectations and share them with your team and manager. Define what's realistic and clarify what's not. For example, if you're expected to "come up with new product ideas that increase revenue," but your focus for the next six months is migrating a monolith to microservices, make that clear. Similarly, if some team members expect you to code as much as they do, explain why that may not be feasible.

The goal is to ensure everyone is aligned and understands your priorities. Setting clear expectations early on helps avoid misunderstandings, disappointment, and potential conflicts down the line.

WHAT IF THERE IS NO JOB DESCRIPTION?

It's rare, but it happens. I like to see this as an opportunity. If there's no job description, create your own by applying the same principles:

- Jot down what you think is expected from your role.

- Ask your stakeholders and team members about their expectations.

- Compile all inputs into a document.

- Share it with your manager and team to get agreement.

An added bonus: you can make this document visible across the company and ask for more input. It might just kick-start the creation of an official job description.

IS THIS ROLE A GOOD FIT FOR ME?

Deciding if the tech lead role is right for you is deeply personal and unique to your situation. No one can answer this question but you. Start by exploring what the role entails and how it aligns with your goals.

Here are some activities that can help you decide:

- Talk to other tech leads about their day-to-day experiences, and see how you feel about what they are sharing. Do you like what you hear?

- Read through your company's expectations for the role. Keep in mind that it varies by company, so explore which environment aligns with your own definition. Don't let anyone fool you: no matter how technical the role looks on paper, there will always be a lot of "dealing with people" in the role.

- Get feedback. Ask people around you how they perceive your fit for the role and why. What strengths do they see that align with it? What might hold you back?

- Talk it over with someone objective. People often bring up this topic to me in coaching. Surprisingly, very often, after just an hour of deep diving into their reasoning and worries, they get clarity on wanting to pursue the role or not.

This being said, the only way to really know is to try it out.

Common Pathways to Becoming a Tech Lead

There are three main pathways that can lead you to a tech lead role. In this section, I'll break them down one by one, sharing insights into how each works, the pros and cons, and how they might align with your own journey.

TAKING OVER THE ROLE ON YOUR CURRENT TEAM

The tech lead of your team announces they're leaving, and you realize this could be your chance. This is exactly how I became a tech lead at Thoughtworks. When my current tech lead announced he was rolling off the project, I immediately offered to step in. It felt like the natural next step for both me and my team, aligning perfectly with my long-term goals, and the timing was perfect.

I had already been preparing to jump into a leadership role for a while. I took part in multiple leadership trainings, and I was taking on more and more responsibilities in my current team and outside of the team that required leadership skills. I was leading big initiatives that involved direct and constant interaction with the clients, growing other people in the company, and becoming more and more visible inside and outside the company by speaking at different events.

Having been on the team for nearly two years, I was the person with the longest tenure on the team (I was in a consultancy environment, where people rotate teams way more often than in a product company), so I had the most context on our products and technical solutions. In addition, I was the most excited candidate and felt confident I could take on the role.

Based on these reasons, it was no surprise when the leadership team said yes to my request, with the condition that I prepare for a smooth transition. Together with my current tech lead, we built a plan that allowed me to step into the role gradually. I initially took on the position of secondary tech lead (the tech lead-in-training, preparing to take over), allowing me to get a feel for the responsibilities while still having support.

Here's how the transition played out over the two months before my tech lead rolled off the project.

We kicked things off by setting up a clear two-month timeline. I started attending meetings with stakeholders and other teams, shadowing my tech lead at first, then gradually taking over and leading them on my own. Each week, we held handover sessions where I got a behind-the-scenes look at the tasks my tech lead was handling and caught up on team initiatives I hadn't been directly involved in before.

I began stepping into decision-making conversations within the team, initially with my tech lead there to guide me. After each session, we'd debrief, and I'd get feedback on what to tweak or improve. I also took over the tough conversations (risks, delays in delivery) with the client, knowing my tech lead would step in if things got tricky. On the internal side, I started handling progress updates and addressing potential issues with stakeholders, again with my tech lead as a safety net if I needed backup.

Looking back, there's one key area I would approach differently if given the chance: ensuring clear communication with the team during the leadership transition. Here's what I wish had happened, and what I actually did:

What I wish had happened

In retrospect, my plan should have included clear communication to the team from my tech lead or someone on the leadership team about the transition, explaining that I would be taking over the role, when it would happen, and outlining the plan for the transition.

What happened

When I transitioned into the tech lead role, there was no official communication from leadership to my team; only the stakeholders were informed. I assumed the team knew, as I became less involved in daily tasks and focused on the handover plan. After my tech lead left, I started leading meetings and setting up one-on-ones. This confused the team, making them feel like I was overstepping.

During a one-on-one, someone finally said, "So, you're the tech lead now? I assumed, but no one confirmed." That's when I realized there had been a lack of clarity. I asked leadership to officially announce my new role, and once they did, along with an apology for the oversight, things fell into place.

This experience taught me the importance of clear communication. If you're transitioning into a leadership role, ensure this step is part of the process. It saves confusion, builds trust, and sets the tone for your leadership journey.

And sometimes, it's the other way around: your team might know, but stakeholders or other supporting functions may be left out. Even when a promotion is announced internally, it's not always obvious across functions. That can lead to confusion about responsibilities, misalignment in collaboration, and delays in support. So, when planning your transition, think about everyone that needs to be in the loop and make sure they are informed.

The benefits of this pathway are as follows: the transition process into the role becomes smoother since you're already familiar with the team, the technology, and the product, and you have your current tech lead by your side to guide your first steps.

The main downside of this approach is that your team members already know you as a fellow developer, and it can take time for them to start seeing you as a leader. I experienced this myself: because they knew me so well, it took longer for them to trust me in a leadership role. People became more hesitant to share everything with me, and at times, my authority wasn't taken as seriously. However, these issues can be managed by setting clear expectations with the team and addressing any concerns early through honest conversations.

There's also a variation of this situation that can be even trickier: when the outgoing tech lead doesn't leave the team but instead moves into an individual contributor (IC) role. While this can be a healthy shift if managed well, it often creates confusion or even tension. The new tech lead may struggle to fully establish authority if the previous lead's presence looms too large, especially if they continue contributing heavily or unintentionally undermine decisions. This inverted power dynamic can make it harder for the new lead to bring change, and requires strong alignment, communication, and mutual respect between the two.

What I'd advise in this case is to sit down with the previous tech lead and align clearly on how you'll move forward together. It can really make a difference if you ask this person to actively support you during the transition, even if your leadership style ends up being different from theirs. This alignment is ongoing work.

Overall, transitioning to being the lead of your current team remains one of the smoothest ways to step into the tech lead role.

TAKING OVER THE ROLE ON ANOTHER TEAM

Sometimes, you spot the opportunity yourself. Maybe another team needs a tech lead, and you proactively put yourself forward. This could be because you feel ready to grow into the role or because you're curious to go through the process and get feedback. You might also bring it up in a conversation with your manager, asking for their support. Taking this kind of initiative is already a sign of leadership, even before you formally step into the role.

Other times, the opportunity comes to you. For example, the tech lead of another team is leaving, and your manager asks you to step in because you're the best fit based on your skills and experience. You accept the challenge.

To make the transition smoother, it often helps to jump in as soon as possible and have some overlap with the current tech lead. This gives you a chance to cocreate a transition plan, absorb context directly, and ease into the role gradually (like I did with my tech lead; more on this in the previous section, "Taking Over the Role on Your Current Team" on page 15). That said, overlap isn't always possible, or even preferable in every situation.

In some cases, stepping in after the previous tech lead has fully left can actually create space for the team to open up more freely, reflect on what wasn't working, and give you a chance to establish your own leadership approach right from the start. The absence of overlap can create a clean slate, but it also means you'll have to work harder to reconstruct context and rebuild trust without that direct handover.

There's no universal right answer here. Both approaches have trade-offs, and what works best will depend on the team's dynamics, the outgoing lead's leadership style, and how much change the team is ready for.

Transitioning to a leadership role in another team does come with a few extra challenges. Beyond just transitioning to tech lead, you also have to go through a full onboarding process with the new team. They've already established their own way of working, and as a new leader, you may face some friction as they adjust to you.

If overlap with the outgoing tech lead isn't possible, the best approach you can take is to start by focusing on building relationships and gathering as much information as possible about the product and team dynamics before you start to rock the boat. Forming a few allies early on can help you access the knowledge you need and get a better sense of how the team operates.

All this makes it a bit trickier than taking over a team you're already familiar with, where you know the people, product, and technical solutions.

The upside is you're walking into a fresh start: new team, new people, new rules. This is a chance to define how you want to be seen as a leader right from the beginning.

BUILDING AND LEADING A BRAND-NEW TEAM

As your company grows, a new team is formed, and it needs a tech lead. Whether you're asked to take on the role or you volunteer, you're starting from scratch. Everything is new, for you and the team. This gives you a unique opportunity to shape the team's culture, processes, and technical direction right from the start.

Pros: Everything is new, not just for you but for everyone involved. This gives you the opportunity to start fresh, define your role as a leader, improve on your previous experiences, and bring people into the team that are aligned with the culture you want to create. Your team members are also likely to be more engaged and curious, eager to contribute and take initiative at this stage.

Cons: Everything is new, and you don't have someone on the team to guide your first steps.

There's a model that can help make sense of this: the Tuckman model, also known as Tuckman's stages of group development, outlines the typical phases teams go through as they form and mature. You can see all the stages illustrated in Figure 1-3.

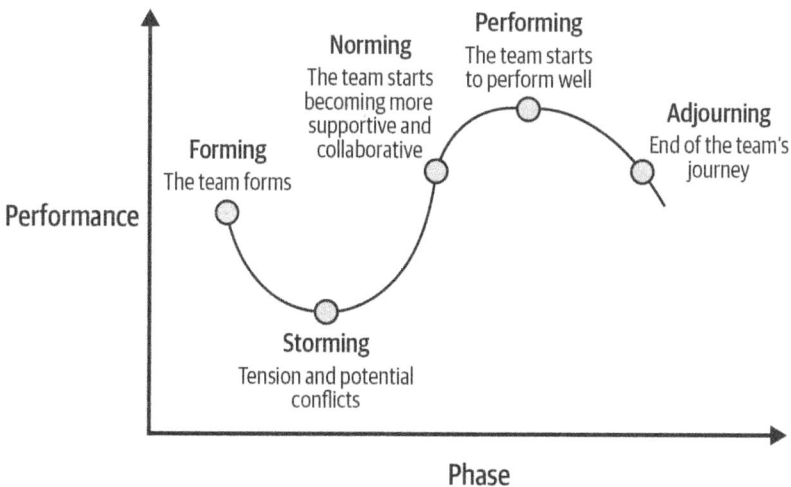

Figure 1-3. Tuckman's team development model

Every new joiner sends a team back to the forming phase, but for a brand-new team this phase comes with added complexity. You're not just onboarding someone into an existing rhythm; you're building that rhythm from scratch. In this setup, the early stages tend to be longer and more uncertain.

You might find yourself moving back and forth between forming, storming, and norming as the team evolves, experiments, and learns how to work together. These stages are rarely linear.

To start with, you'll need to understand what your team will build, why it matters, and how you'll go about it. This often involves aligning with stakeholders, exploring technical options, and finding your team's purpose within the broader organization.

At the same time, you'll need to start shaping the team's culture and establishing ways of working. This can take anywhere from a few weeks to several months, depending on team size, past working relationships, organizational maturity, and how often you actually get to collaborate.

People also need time to get to know each other and find their rhythm. Even experienced engineers need space to build trust and learn how to collaborate effectively.

And as the team's scope evolves, you may need to bring new people on board, adding another layer of transition and adjustment.

These phases don't happen one at a time. They overlap, repeat, and evolve. But knowing they exist helps you recognize where your team is and respond with more clarity, patience, and intention.

Required Skills and Mindset Shifts

As a tech lead, you'll need more than just technical expertise or leadership skills; it's about embracing a new way of thinking.

In this section, I'll first explore the technical skills you'll need, focusing on the importance of understanding the software development lifecycle and how it shapes your team's work. Then, I'll dive into the leadership skills and mindset shifts that will help you empower your team, align with stakeholders, and adapt to the complexities of this role.

TECHNICAL SKILLS

In the section "Defining the Scope" on page 3, I explain why having technical breadth as a tech lead often matters more than being an expert in every technology your team uses. Your goal is to ensure that your team collectively has the capabilities to meet the expectations of the work. Of course, this doesn't mean you can skip technical knowledge entirely; you still need to understand enough to guide your team effectively.

Let's start with an overview of all the different stages in your team's development cycle. I'm a big fan of teams having autonomy over what they build, which is why I embrace the "you build it, you run it" methodology. This approach requires your team to handle every stage of the software development lifecycle (Figure 1-4), from planning and analysis to design, implementation, testing, deployment, and maintenance.

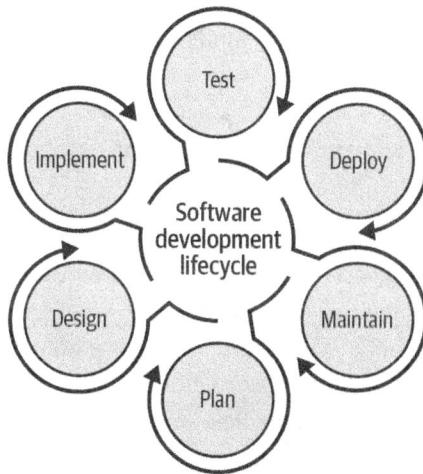

Figure 1-4. Software development lifecycle (SDLC)

When your team owns the full lifecycle, there's no way around it: you need to understand your product at every stage. This requires a solid high-level understanding of the whole process:

Planning, analysis, and design

A big part of your role is planning and analysis. Planning includes prioritizing competing features, estimating effort (as frustrating as that may be), setting milestones, and contributing to roadmap and capacity planning. While you may not own all of these tasks outright, and the responsibility for getting them done is shared across the team, product managers, and stakeholders, you're often expected to guide or influence the process. More on this in the section "Planning and Keeping a Project on Track" on page 291.

Analysis is about understanding what you're building and why, long before any code is written. That could mean working directly with customers or collaborating closely with a product manager to understand requirements. It also involves balancing cross-functional concerns like accessibility, performance, security, and maintainability, as well as making smart build-versus-buy decisions.

Once those foundations are clear, design comes into play. For these early stages, skills like architecture design and understanding how services are designed and scaled are key. The system architecture your team chooses, whether it's microservices, a monolithic structure, or a hybrid model, will have a significant impact on every phase of development. Knowing the benefits and trade-offs of different architectures enables you to guide your team in making informed decisions about scalability, performance, and maintainability.

For example, working with microservices architecture may require a deeper understanding of distributed systems, interservice communication, and how to manage dependencies. Meanwhile, monolithic architecture focuses on internal cohesion and the impact of changes on the entire system. As a tech lead, you should be comfortable discussing these trade-offs with both your team and product stakeholders.

Implementation

During the implementation phase, your focus shifts to development. Depending on your product and the problem your team is solving, you might need stronger backend development skills, such as working with APIs, backend frameworks, database interactions, or database management. For frontend-heavy projects, familiarity with modern frontend technologies, performance optimization, and how the UI interacts with the backend will be needed. You might even need both if your project is full-stack. Of course, you'll also need to understand the programming languages your team uses daily to contribute meaningfully.

Testing

Quality is your responsibility. While you won't write every test, you need to ensure proper strategies are in place (unit, integration, and end-to-end testing) and encourage a quality-first mindset. Testing isn't just about catching bugs; it's about avoiding regressions and keeping technical debt in check. You'll also need to understand how your system interacts with others.

Techniques like contract testing and CFRs (cross-functional requirements) can help avoid ugly surprises when systems integrate (more on this in the section "Continuously Testing" on page 245 and the section "Defining and Managing Cross-Functional Requirements" on page 232).

Deployment

Your team's deployment process says a lot about its efficiency and reliability. You need to understand your deployment pipeline, the infrastructure involved, and which parts your team owns. Whether working with cloud services or on-premises solutions, you'll need to troubleshoot slow delivery pipelines and ensure releases are smooth. Familiarity with concepts like CI/CD (continuous integration/continuous deployment or continuous delivery) is nonnegotiable these days for ensuring fast, automated releases and quick resolution of production issues.

Even if your company has a separate team handling deployment, you still need to understand the process at a high level. This knowledge helps you anticipate how deployment considerations might affect your product and ensures your team delivers a reliable solution.

Maintenance

Building the product is just the start; keeping it running is the real test. In my experience, maintenance often takes up at least half of a team's time, and ironically, established products that require ongoing maintenance are usually the biggest revenue drivers. Poor decisions made earlier in development show up here, and fixing them can cost a fortune. Observability skills, including monitoring, troubleshooting, and debugging, are critical to ensure the system runs smoothly and your team isn't flying blind. You can't improve a system if you don't know what's wrong with it.

Even if your team doesn't directly handle all these stages (e.g., there are separate quality assurance or deployment teams), understanding how they apply to your product is crucial. Too many teams focus solely on the implementation phase, which can cause serious problems. Every phase plays a role in the final result, and as a tech lead, you're accountable for that. It doesn't matter how good your code is if it takes three days to reach production. In some contexts, that might be normal, but in fast-moving teams, that kind of delay can become a serious bottleneck. As a tech lead, you're accountable for the entire delivery process, not just the code.

This might feel overwhelming, but you don't need to master all these skills up front. Much of the learning happens on the job, and different areas will demand focus depending on your project's type and stage. Projects also vary; greenfield projects are about new architectures and innovation, brownfield projects involve navigating legacy systems, and scaling projects focus on reliability and performance.

No tech lead knows everything, and that's fine. The key is to focus on what matters now and learn as you go. Work with your team to create growth opportunities. Run experiments, host hackathons, or set up learning days. Side projects can help, but I prefer learning on the job. If your current role doesn't offer the chance to grow, you can propose a new project or even consider moving to one that aligns with your goals.

LEADERSHIP SKILLS

Most tech leads think they need only strong technical skills in order to be effective as tech leads. I definitely believed so, and I was quickly proven wrong once I stepped into the role because the challenges that I encountered daily required soft skills.

Here are some examples:

Two developers fighting for hours over what JSON parsing library to use
My first instinct, like most tech leads, was to dive into the technical options myself and pick the "best" solution. I definitely made this mistake early on. But over time, I learned to ask questions before jumping into action. By listening to both sides, I realized that, surprisingly, the conflict was not about the parsing library, but actually, these two people had a recurring underlying conflict between them, so they could not agree on pretty much anything. So, instead of diving into the JSON libraries, I helped them resolve their personal conflict.

The team wasn't addressing tech debt
Despite having the knowledge, the time, and the product buy-in to tackle it, nothing was progressing. The root issue wasn't a lack of technical capability but the absence of someone willing to take ownership and drive the process forward. As a tech lead, you could easily step in and take charge, but that's not sustainable long term. The more effective solution is to enable and empower a senior team member to take on this responsibility and lead the initiative.

All these situations, along with countless others I've encountered or seen other tech leads face daily, have led me to the same conclusion:

Most tech problems are people problems.

And solving these people problems requires strong soft skills.

I dive deep into all the soft skills you'll need as a tech lead in the *Soft Skills for Tech Leads* O'Reilly online course and Chapters 3 through 7, but for now, know that there are seven soft skills every tech lead needs:

Listening
> Focus on listening. Everything starts with listening more.

Building relationships
> Building strong relationships will help you set the foundation for collaboration, alignment, and commitment.

Giving feedback
> Feedback is your best tool for growing your team members and yourself.

Delegating
> Delegation is the secret to effective leadership and team empowerment.

Facilitating
> Facilitation will help you bring all voices to the table and help your team reach decisions collaboratively.

Resolving conflicts
> Disagreements are not only inevitable but are actually a sign of a healthy team: you just have to learn how to navigate them effectively.

Mentoring and coaching
> By making use of mentoring and coaching tools, you can help your team grow without telling them what to do.

Investing in your soft skills is always a wise decision. They never become outdated and are universally applicable. After all, no matter how much tech develops, you will always have to work with people, so you might as well get better at it.

MINDSET CHANGES

As a tech lead, you need to shift from an individual contributor mindset to a leadership mindset as your success is now directly tied to the success of your team, not just your individual performance. This transition involves three key mindset changes, as shown in Figure 1-5.

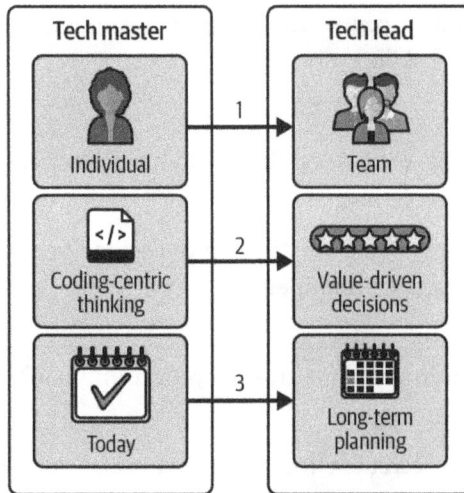

Figure 1-5. Three mindset changes required as a tech lead

The first shift, from individual to team focus, means understanding the following ideas:

- Your technical expertise is irrelevant if your team is constantly struggling.
- The number of tasks you complete doesn't matter if your team isn't delivering at the same pace.
- Finishing your tasks quickly doesn't help if your team is slow to deliver.
- Your self-assessment of doing a great job as a tech lead is meaningless if your team disagrees.
- Personal results don't count if your team isn't achieving success.

- How well you work on your own does not matter if you can't collaborate with your team.

- Talented team members alone aren't enough if they can't collaborate effectively.

As a tech lead, it's about ensuring the whole team moves forward and succeeds together.

Equally important is transitioning from coding-centric thinking to value-driven decisions. Being a tech lead isn't about the amount of code you write but the value your team delivers. Instead of always pushing for the latest technology, prioritize what works best for your team goals and gains team buy-in. Before jumping into coding, ensure assumptions are clarified and strategies aligned. And instead of dismissing meetings as useless, challenge their purpose to make them more effective for everyone.

Then there's the need to step into long-term planning. The decisions you make today will affect your team tomorrow, so you need to understand how to balance immediate tasks with the long-term health of the project. Decisions made for short-term gains should not come at the expense of long-term maintainability or technical debt.

Empowering others becomes a core principle. Instead of defaulting to "I'll handle this," shift to "How can I enable someone else to take ownership of this task?" as the first approach is not scalable. (More on this in Chapter 6.)

Last, letting go of control is essential. As an individual contributor, you had a clear focus: your tasks moved from To Do to Done, often with little interference. But as a tech lead, you're juggling multiple team members' progress and tackling issues that might not even be on the board. Trying to micromanage it all is a one-way ticket to burnout (trust me, I've been there). Learning to trust your team and step back is essential. For more insights, see the section "Avoiding the 'Therapist' Trap" on page 51.

Tip

To help with your mindset shift, write the following sentence down and keep it somewhere visible as a daily reminder: "As a tech lead, you are as successful as your team."

INTEGRATING AI INTO YOUR TEAM

To write this section, I reached out to someone I deeply admire: Birgitta Böckeler (*https://birgitta.info*), a former colleague of mine at Thoughtworks and a true thought leader in this space. Birgitta is a distinguished engineer and currently serves as a subject matter expert and developer advocate for AI-assisted software delivery at Thoughtworks. With over 20 years of experience as a software developer, architect, and technical leader, she brings not only deep technical expertise but also a strong focus on how teams actually work and deliver software together.

Birgitta has been at the forefront of exploring how generative AI impacts software delivery, advocating for responsible experimentation and practical adoption strategies. Her writing and talks offer clarity, realism, and excitement in a field that often gets clouded by hype. I'm incredibly grateful she jumped in to help shape this section with her insights and recent work.

With her help, I've pulled together what I believe every tech lead should understand about AI's role in software development today. Let's jump in.

Note

In this book, I often refer to "AI," "GenAI," and sometimes even "LLMs." While these terms are sometimes used interchangeably in everyday conversations, they aren't exactly the same. Artificial intelligence (AI) is a broad field that covers everything from rule-based systems to self-driving cars. Generative AI refers to models that create new content, including text, code, and images, based on patterns in data. And within GenAI, large language models (LLMs) like ChatGPT, Gemini, or Claude are a specific kind of tool focused on working with language and reasoning over text. Most of the tools discussed here fall under this last category. So when I say "AI" in this book, I usually mean "GenAI," and more specifically "LLMs."

AI is here. It's been here for some time, but now, more than ever, it's everywhere. We might not all be fans of the hype, and it's OK to have mixed feelings about it. But one thing is clear: we can't ignore it. AI is now part of our toolbox. It's useful, it's evolving rapidly, and it's not going away.

Embracing AI thoughtfully is becoming a leadership skill in itself. We need to understand its potential, its limitations, and how it can support the work our teams are doing. This includes considering ethical implications, impact on team dynamics, and long-term maintainability.

Your team will look to you for guidance on how to use it, when to trust it, and when to push back. The challenge is not just learning the tools yourself but helping your team develop safe AI collaboration and good judgment.

Use AI for your own work

Start with your own workflow. Experiment with AI on small, low-risk tasks: generating release notes, cleaning up documentation, summarizing architecture decision records (ADRs), or turning rough notes into onboarding guides. Some teams go further and train models on their internal documentation so new joiners can ask onboarding questions in natural language. Even without that investment, you can use AI as a smarter search layer across your team's existing documents and templates. By trying these applications firsthand, you gain the experience you need to guide others.

Help the team use AI effectively, creatively, and safely

Your leadership is less about being an AI expert and more about setting the culture. Create an environment where your team feels safe experimenting and trying out new tools but is also encouraged to be skeptical, questioning outputs and raising concerns. LLMs don't "know" things; they generate plausible text. They can sound confident and still be wrong. Encourage your team to treat AI outputs like they would advice from an over-confident junior developer: useful but never unquestioned.

To help you help the team, you need to use AI coding assistants yourself to fully understand their strengths and pitfalls. That will enable you to teach your team when to use AI and when to avoid it. You can also ground conversations in context: What problem are we solving? What's at stake if the output is wrong? Build safety nets with strong peer reviews, good test coverage, and knowledge sharing.

Regularly discuss as a team what worked, what didn't, and where AI added value or noise. Social learning is key when it comes to AI.

Amplify your expert knowledge across the team with AI

AI can help you increase your reach as a tech lead. You can curate conventions and rules that reinforce good coding practices and prevent common pitfalls. You can also codify recurring tasks into reusable prompts, including threat modeling, improving ADRs, and structuring requirements, so the AI assistants can amplify consistent approaches.

Beyond prompts and rules, think intentionally about the toolset and MCP (model context protocol) servers you put around the team's AI agents: integrations with other knowledge systems that provide useful context, or secure access to environments for debugging and incident response help. Each of these elements can amplify good practices and increase the probability that AI gives you better results and assistance, but they still don't replace critical review and ongoing validation.

Understand how AI changes learning

The deeper shift is how AI changes the way we and our teams learn. Sometimes the best approach remains doing the work manually, because judgment is forged by wrestling with problems, not bypassing them. Critical thinking, which means questioning assumptions, weighing trade-offs, and understanding context, is more important than ever. Use AI to encourage good habits, like consistent testing or clearer documentation, but don't let it erode the fundamentals. Guide your team to reflect on where AI helps, where it hinders, and what it teaches them along the way.

Manage stakeholders' expectations

AI adoption often tempts organizations to chase numbers: faster delivery, more output, reduced costs. Some executives even assume teams will become "50% faster" just by adding AI to their workflows. These expectations are dangerous. They create pressure to cut corners, skip tests, or compromise quality just to hit inflated targets.

As a lead, your role is to push back against oversimplified metrics and broaden the conversation. Monitor where AI genuinely reduces friction or accelerates learning, but balance that against things like code quality, maintainability, debugging time, and team satisfaction.

Like any tool, the value of AI depends on how thoughtfully it's used.

By setting realistic expectations and widening the definition of success, you protect your team from unrealistic pressure and ensure AI adoption strengthens your outcomes, not weakens them.

> **Tip**
>
> For a further deep dive into the topic, I totally recommend following Birgitta Böckeler on LinkedIn (*https://oreil.ly/yC8u2*). She's constantly publishing insightful articles and sharing up-to-date, practical information on how AI impacts software delivery.

Key Takeaway

The tech lead role can feel a bit vague, and that's often what makes it so intimidating at first. You're expected to lead, but the path isn't always clear. In this chapter, I wanted to bring a bit more clarity and calm to that confusion.

We looked at what the role typically involves and how it can vary across different teams and companies. This ambiguity isn't necessarily a bad thing. It gives you the freedom to shape the role around your strengths, your leadership style, and what your team actually needs.

There's no one-size-fits-all version of being a tech lead, and that's the beauty of it. You can define what success looks like for you, as long as your team's needs stay at the center.

If the role still feels a little daunting, that's OK. It's a big step, but you don't have to have it all figured out on day one. Start small, stay curious, and keep experimenting. That's how you grow into it. That's how you lead.

How to Become a Tech Lead

Becoming a tech lead is an exciting but challenging transition. This chapter will guide you through the journey of stepping into the role, building the skills you need, and overcoming common obstacles along the way.

First, I'll explore how to set a strong foundation for your growth by adopting the right mindset and understanding the importance of intentional development. Then, I'll walk through creating and implementing a growth plan tailored to your unique context and goals. Finally, I'll emphasize the importance of reflecting on your progress and making adjustments to stay aligned with your objectives.

Once your growth strategy is in place, I'll address common challenges that new tech leads face. Time management often becomes difficult to balance, so I'll explore how to effectively plan and manage your new responsibilities. I'll also cover strategies for avoiding the "therapist trap," where you might feel pressured to solve everyone's problems, and how to steer clear of micromanaging your team. Finally, I'll dive into finding the right balance between hands-on coding and providing technical oversight, a delicate but essential aspect of the tech lead role.

By the end of this chapter, you'll have a practical framework for growing into the role of a tech lead and strategies for navigating the challenges that come with it.

Setting the Foundation for Growth

Before you can grow with intention, you need a solid foundation. That starts with how you think, what you understand about your role, and where you currently stand.

It begins with mindset. Embracing a growth mindset means seeing challenges as opportunities to improve and understanding that mistakes are part of the learning process, not something to fear or avoid.

But mindset alone isn't enough. You also need clarity. What's expected of you? What does success look like in your team, your company, your context? By talking with your manager, your stakeholders, and even your peers, you can start to shape a more realistic and grounded picture of what to aim for.

And once that picture is clearer, you can take a good, honest look at where you're starting from. What are your strengths? Where are your gaps? This kind of self-assessment isn't always easy, but it's essential if you want to grow with purpose rather than just reacting to whatever comes next.

These early steps, including how you think, what you understand, and how you reflect, will shape everything that follows. Let's get into it.

BUILD A GROWTH MINDSET

Having a growth plan is effective only if you pair it with the right mindset. To grow as a tech lead, you need to embrace a growth mindset, which is the belief that your abilities can be developed through effort, learning, and perseverance. It's about seeing challenges as opportunities and mistakes as learning experiences, rather than failures.

When I first stepped into the tech lead role, I believed I had to have all the answers and get everything right. I felt like every mistake or moment of uncertainty would undermine my capability in the eyes of others. I would hide my mistakes, thinking it made me appear more competent. I'd prepare technical solutions in advance, rushing to respond to every client question as if hesitation would make me seem less qualified. My anxiety drove me to over-prepare for every situation, leaving me constantly stressed and exhausted.

When others didn't have answers, I felt it was my responsibility to step in and "fix it all," often taking on tasks that weren't mine to handle. Instead of asking for clarification, I'd spend hours figuring things out alone, convinced that admitting I didn't know something would damage my credibility.

Looking back, I realize how limiting this mindset was, because growth requires a fundamental belief: you don't need to have all the answers right away, but you do need to believe in your ability to learn, adapt, and improve:

In order to grow, you need to believe you need to.

The good news is, a growth mindset can be developed just like any other skill. Two strategies helped me transform my approach to challenges and uncertainty.

The first is embracing the phrase "I don't know." Admitting when you don't have all the answers can be a game-changer. It allows you to learn faster by seeking information instead of pretending, reduces the stress of trying to appear all-knowing, and supports a culture of collaboration by making it safe for others to admit what they don't know as well. At first, saying "I don't know" felt uncomfortable; I even got strange looks from others. But over time, it became easier, and I began to see how it strengthened my interactions and trust within the team.

I know this can feel intimidating, and in some environments, even more so. It's especially tough in client-facing situations or in cultures where vulnerability might be viewed as a weakness. I still remember the first time I said "I don't know" in front of a client. It felt ten times harder than saying it to my team. But over time, I saw the same positive effect: it created space for honesty and showed that not knowing is normal. Eventually, more people began to follow. Someone has to be the first to have the courage to say it out loud.

The second strategy is shifting from "I don't know" to "I don't know yet." Reflecting on past challenges often reveals a pattern: you didn't always know the answers, but you figured them out. Think back to moments when you started and you lacked clarity or certainty. What steps did you take to move forward? How did you overcome the fear of not succeeding? Over time, what initially felt impossible became second nature. This mindset shift focuses on the journey of learning and emphasizes that progress takes time and effort.

When it comes to growth, it's also helpful to remember that confidence doesn't always reflect competence. The Dunning-Kruger effect describes how people with limited experience tend to overestimate their abilities, while those who are more skilled often underestimate themselves. If you're doubting your abilities, it might actually be a sign that you're becoming more aware of what mastery really involves, which is a key step in growing. See Figure 2-1 for a visual of the Dunning-Kruger curve, which illustrates how confidence often dips as awareness increases before rising again with true competence.

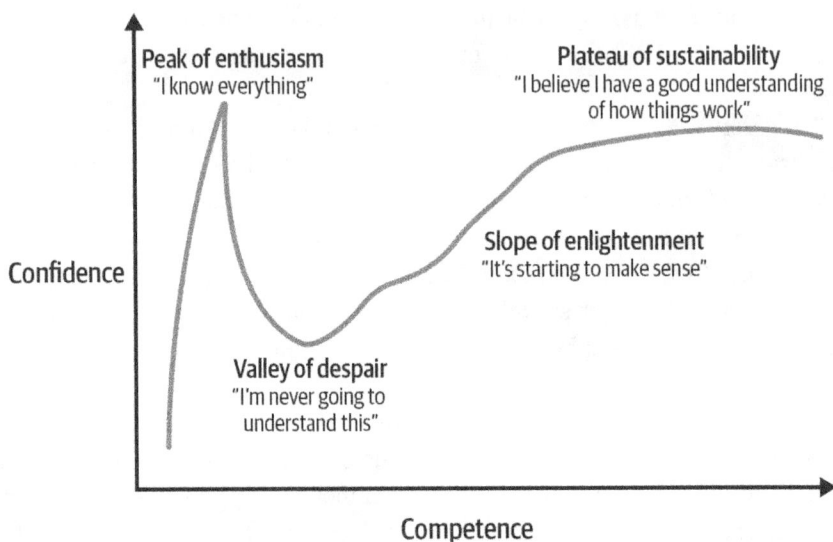

Figure 2-1. *The Dunning-Kruger effect*

It's easy to feel the pressure to know everything, but real growth happens when you trust your ability to learn and adapt. Confidence doesn't come from always having the answers; it comes from recognizing what you know now and believing you'll figure out the rest. Growth starts with this belief. You won't know everything right away, but with time and effort, you will.

Tip

For a deeper exploration of the growth mindset, I highly recommend *Mindset: The New Psychology of Success* (*https://oreil.ly/ZgAcS*) by Carol Dweck (Random House).

CLARIFY EXPECTATIONS

The second requirement for growth is understanding what's expected of you. The tech lead role can vary widely between companies, so it's important to clarify expectations in your specific environment.

One of the most common surprises I hear in coaching is from tech leads who thought they were expected to focus on technical delivery, only to later realize their manager cared more about team development and cross-functional collaboration.

For example, one coaching client stepped into a tech lead role assuming she needed to drive architectural decisions and write the hardest parts of the code. But after a few months, she received feedback that she wasn't visible enough to the team and wasn't spending enough time on mentoring or stakeholder communication.

This situation could have been avoided by clarifying expectations early on. The section "Understanding the Expectations of Your Role" on page 12 offers practical strategies to help you do just that, so you can align your efforts with what your organization actually values from the start.

With these expectations in mind, the next step is to figure out where you stand in relation to them.

ASSESS YOUR STARTING POINT

Once you've clarified the expectations for your role, you can use them to get a clearer picture of where you stand right now. They provide a concrete reference point to define your baseline. A simple way to do this is by using a 1-to-10 scaling system (where 1 means "just starting out" and 10 means "strong in this area"), shown in Table 2-1.

Table 2-1. Assessing your starting point

Step	Description	Reflection Prompt
Rate yourself	Review the expectations identified in the previous section, "Clarify Expectations" on page 36. Rate yourself from 1 to 10 for each area. Use examples. Pay attention not just to what's missing but also to what you're already strong at.	Why this rating? Why not a 10? What's missing to get to a 10?
Seek feedback	Ask colleagues and your manager to rate you on the same scale. Request specific examples to support their rating and suggestions for improvement.	What am I already doing well in this area? What's one thing I could do to improve this area?
Compare and align	Look at how your self-ratings compare to the feedback. Identify gaps and strengths to inform your growth plan.	Where are the biggest gaps? What surprised you?

The main reason why I like this framework is that it forces people to actually evaluate where they are. It's not about the number but about the reflection it triggers. In my experience, people tend to undervalue themselves when self-rating. Many of my coaching clients say things like "I could never rate myself a 10; that just feels like bragging," even when they can't identify a single improvement point in that area. You have to consider that if nothing's missing, maybe you're already a 10.

It's a shame to automatically downgrade yourself like that. If you don't stand up for yourself, who will? Let others point out the gaps. Your job is to own your strengths with clarity and confidence.

Developing a Personal Growth Plan

People often get trapped in the cycle of "doing" without a clear destination, which can drain motivation. I've been there, focusing on tasks without knowing exactly where they're leading. A growth plan changes that by being intentional about where you want to go. It gives you control over your career path, helping you recognize opportunities as they come and make decisions more easily because you know your "why." It's about moving with purpose rather than drifting on someone else's agenda.

The five steps for building a growth plan are as follows.

STEP 1: DEFINE A CLEAR GOAL

A lot of people struggle to define meaningful career goals, especially when transitioning from individual contributor to leadership roles. It's not always easy to articulate what you want beyond a title like "tech lead." That's why it helps to go deeper and get specific.

Ask yourself the following questions:

- What would success look like in this role?
- How will I know I've reached my goal?
- What will be different in my day-to-day work?

But don't stop there. One of the most important questions I ask people when they're seeking a promotion is this: Why do you want it? Is it about recognition, influence, salary, visibility, or something else? Most people's answers are different, and knowing yours can help ensure the goal is truly aligned with your deeper motivation. Sometimes, the promotion you're chasing isn't the only (or best) way to get what you're after.

So instead of just saying, "I want to become a tech lead," try defining what that means to you:

- Leading a team of five engineers on a customer-facing product
- Taking over the tech lead role in my current team when my tech lead leaves
- Having the official tech lead title

The more specific you are, the easier it is to use this goal.

Tip

If you're unsure, look at job descriptions or company expectations, or talk to current tech leads. Ask them: "What made you feel like you were finally in the role?"

STEP 2: DEFINE THE STEPS TO GET THERE

Once you've defined your goal, the next challenge is turning it into action. This is where many people get stuck: they know what they want but aren't sure how to start moving toward it.

Break the goal into milestones: think of them as checkpoints on the path to your destination. These are not just tasks but progress markers that show you're moving in the right direction.

Ask yourself the following questions:

- What needs to be true for me to reach this goal?
- What skills or experiences am I currently missing?
- What small steps can I take in the next week or month?

If the path feels fuzzy, back into it by identifying one or two immediate opportunities already available to you. You don't need a full roadmap, just enough clarity to start walking. You can figure out steps along the way also.

Tip

Review past feedback or performance reviews. Often, they contain useful clues about what you could be doing more of or be doing differently.

STEP 3: BUILD A TIMELINE

Now that you have a goal and a set of steps, it's time to add some structure. A timeline gives your plan momentum. It turns vague intentions into commitments and helps you hold yourself accountable.

This doesn't mean everything has to be perfectly scheduled. The idea is to set rough timeframes for each step or milestone so you can track progress and avoid stalling out.

Ask yourself the following:

- What can I realistically work on this quarter?

- Are there natural checkpoints I can use, like a performance review, team change, or project cycle?

- Which steps depend on others, and how long might those take?

Use a calendar, spreadsheet, or project board (like Trello or Notion) to map it out visually. Seeing it laid out can make the plan feel more real and manageable.

Tip

If you're unsure about timelines, start with short-term goals and adjust as you learn more. Even a rough guess is better than none; it gives you a direction and makes it easier to reflect and improve later.

STEP 4: BUILD A SUPPORT NETWORK

One of the most overlooked, but powerful, parts of a growth plan is asking, "Who can help me?"

Growth doesn't happen in isolation. The people around you—your manager, your current tech lead, mentors, peers, or even a coach—can play a huge role in how fast and smoothly you progress.

Different people can support you in different ways: your manager can give visibility into future opportunities, suggest growth areas, and connect your goals to team needs; your current tech lead might let you shadow them or gradually take over rituals like planning or standups; mentors or peers can share how they approached similar transitions and offer perspective from outside your team; a coach can help you reflect, challenge your assumptions, and stay focused.

Let them in on your goals. When people know what you're aiming for, they can offer better feedback, suggest relevant opportunities, or support you when you need a push. Be proactive: ask for a regular check-in, share updates on your progress, and ask for help identifying blind spots.

This isn't about asking people to "do it for you" but about building a circle of support that helps you grow faster and more sustainably.

STEP 5: IDENTIFY OPPORTUNITIES AND RESOURCES

A solid plan also takes advantage of what's already around you. The question here is "What opportunities and resources can I tap into to help me grow?"

Start with your current role. There are likely more learning opportunities than you realize; you just need to spot them or ask for them. Look for growth moments like leading a team retrospective, even if informally; taking the lead on a small project or feature; facilitating technical discussions; mentoring a more junior team member; or owning a process improvement (like refining your team's release workflow).

Ask your manager or tech lead, "Is there an area where I could take more ownership or help the team?" You'd be surprised how often the answer is yes.

Next, look outside your day-to-day tasks: shadowing others (e.g., sitting in on planning meetings, architecture reviews), contributing to cross-team initiatives or guilds, presenting at internal knowledge-sharing sessions, or writing technical documentation or proposals.

And don't forget about formal resources, like your learning and development (L&D) budget. Use them for leadership courses, workshops, coaching, books, podcasts, communities, internal or external leadership programs, or training opportunities.

To make this more tangible, let's walk through how these five steps come together in practice. The example in Table 2-2 outlines a growth plan for someone aiming to take over the tech lead role in their current team when their tech lead leaves.

Table 2-2. An example growth plan

Step	Example
Define a clear goal	• Become the tech lead of my current team
Define the steps to get there	• Inform current tech lead and manager about my intention • Shadow current tech lead during strategy planning • Lead the migration initiative in the team
Build a timeline	*Lead the migration initiative in the team* Feb — March — April — May — June — July *Shadow current tech lead during strategy planning*
Build a support network	• Schedule one-on-ones with manager • Ask Ana (a tech lead from another team) to be my mentor • Join the monthly company "tech lead conversations"
Identify opportunities and resources	• Use L&D budget for leadership course (be specific) • Read *Leveling Up as a Tech Lead* • Take the *Soft Skills for Tech Leads* O'Reilly online course • Read *The Manager's Path* by Camille Fournier • Join the LeadDev community (*https://leaddev.com*)

The plan may not always go as expected, but it will get you to your goal faster than just going with the flow. Keep adjusting as you progress.

Implementing the Growth Plan

This is the part where your growth plan comes to life.

You've defined your big goals; now it's time to put them aside for a moment and start zooming in. Focus on the small, concrete steps you outlined in your growth plan. Start acting on them one by one while constantly reflecting on what's working and what isn't.

REFLECT

Reflection is a tool that vastly accelerates your growth. For each step you take, assess whether you're aligned with your growth plan. Ask yourself, "Am I making progress?" and "What adjustments are necessary?" Use your milestones as checkpoints to assess whether your actions still support your bigger direction.

A simple way to evaluate your progress is to repeat the same 1-to-10 scaling system described in the section "Assess Your Starting Point" on page 37. Revisit your ratings, ask for updated feedback from others, and have recurring conversations with your team, manager, or peers about your development. These insights can validate your progress or surface areas that need more attention.

Getting honest, useful feedback is harder than it seems. People might hesitate to be direct, especially when you're in a position of authority. That's why it's important to create psychological safety, ask thoughtful questions, and normalize feedback as a regular part of working together. Chapter 5 includes the section "How to Get Useful Feedback from Your Team", which dives into specific strategies you can use to gather meaningful input from both team members and stakeholders.

TRACK YOUR PROGRESS

Whatever comes up at each reflection point, track it. It sounds so simple, but this critical step is often dismissed as "obvious" and skipped entirely. The result? You lose sight of how far you've come and the progress you've made.

You can track your progress in whatever format works best for you. Some people prefer a handwritten journal or notebook, while others use a spreadsheet, a Notion page, or a digital app like Trello or Todoist. The key is to make your progress visible. A simple table with your goals, key steps, timeline, and a column for status/notes or reflections (like in Table 2-3) can go a long way.

Table 2-3. Progress tracker

Goal	Step	Timeline	Status/Notes
Improve collaboration in the team	Pair program with three different team members	By August 1st	June 3rd: paired with one team member
	Have two other team members pair together on a task	By August 1st	June 3rd: the plan is for them to pair next week
	Have two team knowledge-sharing sessions	By August 1st	June 15th: I delivered the first session; encouraged another team member to lead next

You can be as specific as you like or keep it extremely lightweight. In fact, the simpler your system, the more likely you are to stick with it. It could be as easy as keeping a running list in a notes app where you jot down small wins as they happen. For example: "June 3rd: Convinced Alex to pair with me on the new onboarding flow, and it went really smoothly." You can later match these small wins to the goals you've defined for yourself, which is especially useful during performance review cycles, where clear, goal-aligned examples go a long way.

Many people avoid tracking progress for reasons that might seem valid at first.

Some feel that being structured about their growth is selfish or self-centered, something others might frown upon. I constantly have to remind people that there's nothing wrong with wanting to grow and become a better person or professional.

Others hesitate, asking, "What counts as progress?" The answer is straight-forward: anything, big or small, that moves you forward is worth recording.

Then there's the classic excuse: "I'll remember it." The reality is, you probably won't, just like many tech leads I talk to daily. And why overburden your brain when writing things down is far easier and more effective? Writing things down activates a different part of your brain, making your progress feel tangible and keeping you connected to the steps you've taken. These small habits, simple as they may seem, can fundamentally shift how you perceive and maintain your growth.

You don't need to track daily, but finding a regular rhythm that works for you, whether it's weekly, biweekly, or monthly, can make a big difference. The key is consistency.

Tracking offers two significant benefits. First, it boosts your self-confidence. On tough days, looking back at the challenges you've overcome reminds you of how capable you are. Second, it provides motivation. Progress can feel painfully slow when you're focused only on the long road ahead, but reflecting on how far you've come can restart your drive to keep going.

Tip

Don't skip reflection. It helps you stay motivated and on track.

ADJUST

Growth isn't static; it's a continuous process of learning and adapting. Use the feedback and insights you've gathered to refine your next steps.

Start by being intentional about what you focus on next. If you want to work on your communication skills, lean into people interaction and find opportunities to practice. If there's a technical area you want to deepen your knowledge in, prioritize pairing with experienced team members and working on tasks that stretch you. This is how you take control over your growth instead of just going with the flow.

Be strategic about your environment. If you can choose your projects, pick ones that help diversify your experience. But even within your current role, there are plenty of opportunities to grow; you just have to be intentional about spotting them.

By implementing, reflecting, tracking progress, and adjusting, you'll turn your growth plan into a dynamic tool that evolves with your goals and context, keeping you on track to becoming an effective tech lead.

Overcoming Common Initial Challenges

When I talk to tech leads, I see the same struggles coming up time and time again. Some feel overwhelmed trying to manage their time effectively and balance their focus among technical responsibilities, team dynamics, and stakeholder needs. Others fall into the trap of trying to solve everyone's problems or micromanaging their team. These challenges are not just common; they're expected. But they don't have to derail you.

In the sections that follow, I'll walk you through these recurring challenges one by one and share strategies for navigating them effectively, starting with time management.

TIME MANAGEMENT

In the section "Day-to-day Responsibilities" on page 10, I outlined the core daily tasks for a tech lead: meetings, coding, and thinking time. However, no matter how well you plan, disruptions are inevitable, whether it's production incidents, strategy changes, or team conflicts. You need to be prepared to switch lanes and adapt.

While flexibility is key, it's still important to have a time management system in place. Time management is a major challenge for most tech leads, and this section will help you build a process to gain control over your time and handle disruptions effectively.

Step 1: Visualize your workload

Visualizing your workload can help you spot priorities, manage your time more effectively, and reduce overwhelm. There are many ways to do this, and different formats work better for different people.

You might prefer using a color-coded task list to quickly flag urgent and non-urgent work. Mind maps can help you see how tasks relate to each other and to broader goals. Kanban boards (physical or digital) are great for visualizing progress and balancing workload. Some people use weekly planning templates or calendar blocking to map time against task types.

There's no one best method; it's about what helps you stay focused and intentional.

As an example, I'm using the Eisenhower Matrix here to show you how one approach can help you sort tasks by urgency and importance. In Figure 2-2, you'll find examples of what types of tasks fit into each quadrant.

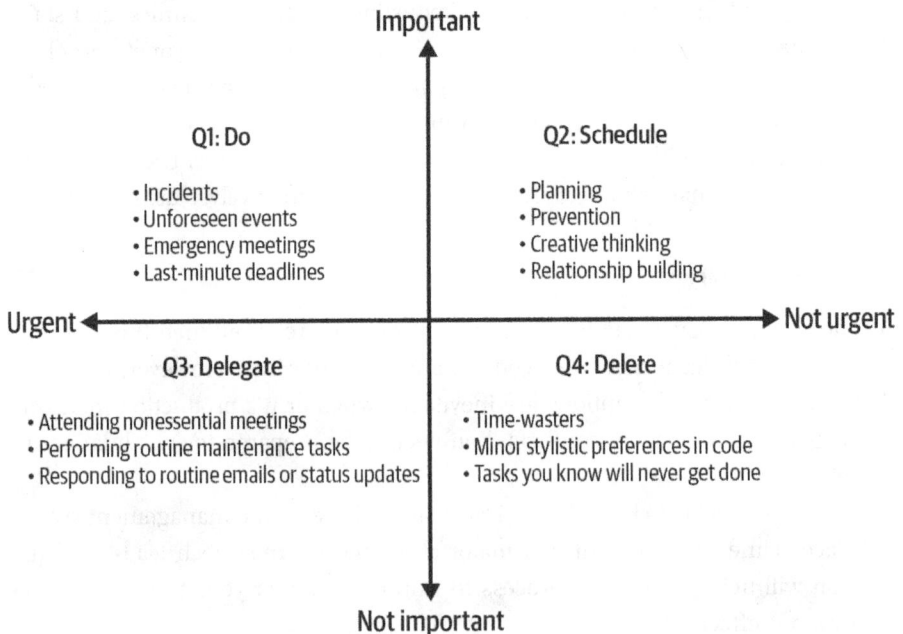

Important

Q1: Do
- Incidents
- Unforeseen events
- Emergency meetings
- Last-minute deadlines

Q2: Schedule
- Planning
- Prevention
- Creative thinking
- Relationship building

Urgent ← → **Not urgent**

Q3: Delegate
- Attending nonessential meetings
- Performing routine maintenance tasks
- Responding to routine emails or status updates

Q4: Delete
- Time-wasters
- Minor stylistic preferences in code
- Tasks you know will never get done

Not important

Figure 2-2. The Eisenhower Matrix—task types example

Let's take a practical example: a typical first attempt by a tech lead to populate the Eisenhower Matrix, as shown in Figure 2-3.

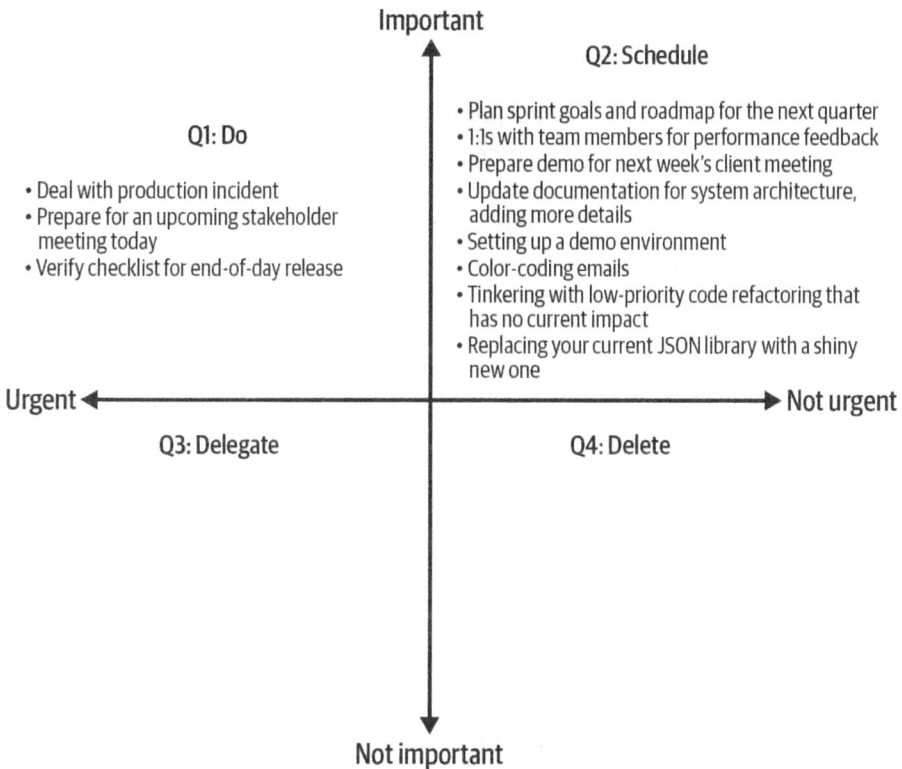

Figure 2-3. The Eisenhower Matrix—filled in

Step 2: Analyze your task distribution and adjust your workload

When you look at your task list, there are a few things to keep in mind. First, take a moment to step back and ask yourself: do you have tasks in all four quadrants? One of the most common struggles for tech leads during this exercise is that they fill in only the top two quadrants, as you can see in Figure 2-3. When asked why, the answer is almost always the same: "Everything is important; that's why it's on my list." This is a classic example of the urgency effect, a cognitive bias where we tend to prioritize tasks that appear urgent, often at the expense of more important but less time-sensitive work. Just because something feels immediate doesn't mean it deserves your time or energy.

Let's take a closer look at the board and see what might be moved to the lower quadrants.

Think about the tasks that truly require your unique expertise as a tech lead. Are they things that only you can handle? Some tasks clearly fall into this category, things like showing up for one-on-ones with your team members or joining high-level stakeholder conversations where you're the one who holds the full context. But then there are others. Preparing a demo for a client meeting next week or setting up a demo environment might initially feel like your responsibility, but these are tasks that someone else on your team could take over. By delegating them, you free yourself to focus on what only you can do, and you give others on your team a chance to shine and grow.

Now consider another type of task: the ones you're doing just because it would be nice to have them done. For example, adding an extra level of detail to your system architecture documentation. Sure, you could argue that this is important. But why? Is another team depending on it? Will the product stop working if it's not done? Or is it just that you want everything to be perfectly up-to-date? If the answer to the first two questions is no, and the third one resonates a bit too much, then this is exactly the type of task you can let go of. Move it to the Delegate or Delete quadrant. Let someone else handle it, or let it go altogether.

What happens if I don't do this? This is the key question to ask when looking at tasks like "Color-coding emails," "Tinkering with low-priority code refactoring that has no current impact," or "Replacing your current JSON library with a shiny new one." The honest answer is: probably nothing. These are "nice-to-haves" that often sit on your to-do list but don't add real value. They don't significantly impact your work or your team, and holding onto them just takes up mental space. Move them to the Delete quadrant. If one of these tasks turns out to be really important, it will come back up when it's relevant. Until then, let it go and focus on what matters most.

Not having all the quadrants filled in might also be a sign that you're getting stuck in one type of activity, losing track of others, or overlooking blind spots. These blind spots can manifest as areas where you're wasting effort, missing opportunities, or spreading yourself too thin.

If you're spending most of your time in Q1 (Important & Urgent), it's a clear indicator that you're in constant firefighting mode. Things keep popping up unexpectedly, leaving you feeling off guard and reactive. This often means you're not investing enough time in Q2 (Important & Not Urgent), planning and focusing on long-term goals. If you're unsure what to include in Q2, ask your team and your stakeholders: "What should I be focusing on that I'm not?" Their perspective can help uncover opportunities you've overlooked.

On the other hand, spending too much time in Q3 (Urgent & Not Important) suggests you're being distracted by tasks that don't need your direct involvement. These tasks often feel pressing but offer little value. Instead, consider delegating them to your team.

Finally, if you find yourself stuck in Q4 (Not Urgent & Not Important), it's time to reassess. Low-impact tasks might be your way of avoiding more significant challenges. Ask yourself: are you procrastinating on something more meaningful? If so, don't hesitate to reach out for support, whether from your team or peers, and redirect your focus to what's most important.

By confronting these tendencies, you can move past distractions and focus your energy where it really counts.

Based on this analysis, the board in Figure 2-4 is now much more balanced.

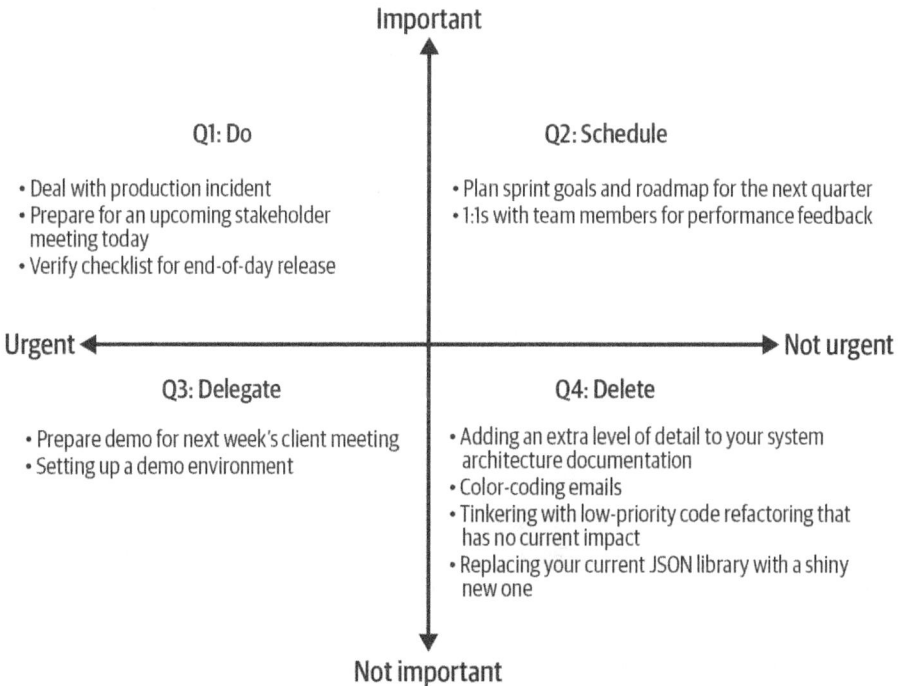

Important

Q1: Do

• Deal with production incident
• Prepare for an upcoming stakeholder meeting today
• Verify checklist for end-of-day release

Q2: Schedule

• Plan sprint goals and roadmap for the next quarter
• 1:1s with team members for performance feedback

Urgent ← → **Not urgent**

Q3: Delegate

• Prepare demo for next week's client meeting
• Setting up a demo environment

Q4: Delete

• Adding an extra level of detail to your system architecture documentation
• Color-coding emails
• Tinkering with low-priority code refactoring that has no current impact
• Replacing your current JSON library with a shiny new one

Not important

Figure 2-4. The Eisenhower Matrix—all quadrants filled in

While it's not necessary to have tasks in all quadrants, the exercise of filling them all forces you to evaluate the true value of each task and prioritize more effectively. The reality is that you'll likely never finish everything on your to-do list, so making peace with that and learning to delete or let go of less important tasks can help you feel more productive. By clearing away the unnecessary, you'll free yourself from the weight of an overwhelming to-do list and gain the focus needed to concentrate on what matters.

Step 3: Plan your week

Once your matrix is filled in, the next step is to extract your key tasks for the upcoming week and map them to specific time slots in your calendar. This is where the magic happens, turning your priorities into a realistic, actionable plan.

Start by asking yourself: does everything fit within your schedule? If not, it's time to reevaluate. Is there something that can be removed or delegated? Be honest about what truly needs your attention and what doesn't. Don't forget to include breaks and some realistic buffer time to account for the inevitable unplanned disruptions that creep in.

Planning your week this way gives you a structured approach to your time. It allows you to focus on what matters most and avoid falling into the trap of prioritizing based on "who screams the loudest." Instead of feeling overwhelmed, you'll feel in control, with a clear plan to tackle the week ahead.

Step 4: Review and adjust

Reviewing and adjusting your process might feel like just another task on your already overwhelming to-do list, but it's one worth prioritizing. Taking this step helps you uncover patterns, avoid repeating mistakes, and refine a system that becomes more effective over time.

At the end of the week, set aside a few moments to reflect. Consider what urgent issues derailed your plans and whether they were unavoidable. Look at what got deprioritized and ask yourself why. Was it because it wasn't important, or did something else take precedence? Most importantly, think about what you can improve for next week. How much of what you planned did you actually accomplish? If it's around 50%, that's still a solid start. It takes time to get good at planning. Each week gives you a chance to learn and refine your process: Were your estimates off? Did you forget to build in breaks? Did you try to take on too much? Use those insights to tweak your plan for the next week and make it more realistic.

Time management tips

There are many different time management strategies out there that can help you manage your time as a tech lead, but here are three that I find most effective:

time blocking, task batching, and giving myself time to consider whether I can take on a particular task.

The first one is time blocking, a simple yet powerful time management method where you allocate specific blocks of time on your calendar for particular tasks or activities. It's about being intentional with your time, ensuring you create space for what matters most, instead of letting your day get hijacked by unplanned distractions. For instance, if you're pairing with team members daily, you can block dedicated hours for pair programming to keep a consistent routine. The same goes for planning: setting aside focus blocks ensures you have uninterrupted time to step back and think strategically. As a tech lead, your team looks to you for direction and long-term planning, not just hands-on coding, so it's crucial to block out time for this.

Another effective technique is task batching, which works hand-in-hand with time blocking. Grouping similar tasks, like meetings or feedback sessions, into one block reduces context switching and helps you stay focused.

And here's one of my golden rules that helps me protect my schedule: never say yes on the spot. Instead, respond with "Let me think about it," check your calendar, and make sure it fits.

Time management is about making space for what matters most. By setting weekly intentions, visualizing your workload, prioritizing deliberately, and regularly reviewing and adjusting your plan, you create a system that puts you back in control.

AVOIDING THE "THERAPIST" TRAP

One of the most common traps tech leads fall into is trying to solve everyone's problems. Many get into leadership roles because they genuinely want to help others, but this can quickly turn into taking on too much responsibility. What starts with "How can I help?" often turns into "Don't worry, I'll do it for you."

Note

It's not your job as a tech lead to solve everyone's problems!

The pitfalls of over-involvement are hard to ignore. Burnout is inevitable if you're constantly stepping in to handle everything (been there, done that). You'll exhaust yourself trying to manage tasks the team should own. Frustration builds quickly too, together with resentment toward the team for not stepping up, toward stakeholders for endless demands, and toward yourself for not setting boundaries.

Beyond the individual cost, there's also a hidden team impact: when you always step in, your team misses the chance to develop autonomy. People learn by doing (and struggling) through challenges. If you keep solving problems for them, they won't get the practice or confidence to do it themselves.

And then there's the blame. If you're always the fixer, you become the first to blame when things go wrong. Instead of encouraging shared accountability, you've positioned yourself as the only point of failure.

Early in my career, I fell into this trap when a team member, new to the city, shared his frustrations during a one-on-one. He was struggling to sort out his paperwork, and it was affecting his focus at work. Without even being asked, I jumped in, eager to help. "Don't worry, I'll handle it," I said, even though I wasn't an expert in the process myself. I took full responsibility: researching the steps, preparing his documents, asking others for advice, and even sitting with him to book his appointments. I even skipped standups to accompany him to appointments.

It didn't go as planned. We hit roadblocks, and he became increasingly frustrated with me because I hadn't solved it as promised. Meanwhile, I was overwhelmed, my work was suffering, and I felt unappreciated for my efforts. A colleague finally asked me, "Why are you trying so hard to fix his problem? It's not yours to solve." That moment was a wake-up call. I stepped back, pointed him to someone better suited to help, and let him take over the process. But the damage was done; our professional relationship had soured, and both of us were left feeling frustrated.

In hindsight, there were countless moments when I could have handled it differently. Here's how I wish I had approached the situation and how I now tackle similar challenges:

Set proper boundaries

Before diving into any problem, I take a moment to analyze. In this case, I could've simply said, "I'm sorry to hear that" and taken time to reflect. Was it my problem to solve? No; in fact, he hadn't even explicitly asked for my help. Even if someone does, I now ask myself, "What happens if I say no?" Often, the answer is nothing catastrophic, and the person can handle it with some guidance.

Keep yourself in check

Reflection is key. Talking it through with a mentor or colleague can offer valuable perspective. Had I done this earlier, I would've realized I was overextending and could've stepped back. I now routinely ask myself, "Are there certain people or tasks draining more of my energy?" or "Am I stepping in when I should be encouraging ownership?"

Shift your support

Helping doesn't always mean solving. Sometimes it's enough to simply listen: "I'm sorry you're dealing with this." In other cases, the best move is to point them to someone else. When I finally realized the paperwork wasn't my problem, I identified someone better equipped to help him and sent him their way.

The main lesson you can take from this story is to pause before jumping in and ask *Is this problem mine to solve?*

That one question can help you set better boundaries, reduce unnecessary stress, and give the kind of support that's actually needed.

These lessons apply directly to day-to-day team dynamics. For example, if a team member says, "I can't deal with this task; can you take it over?" my instinct used to be to step in and help. Now, I pause and ask what's making it difficult and how they might approach it differently. Often, with a bit of coaching, they're able to handle it themselves and grow from the experience.

It's the same when someone brings me an urgent matter. Before jumping in, I try to understand how urgent it really is by asking, "When do you need an answer?" This gives them a chance to reflect on what they really need and helps you prioritize more intentionally. Often, the request isn't as urgent as it first appears, and they're happy to wait. Sometimes, they might even figure it out themselves before you get to it. But if it is urgent, you can decide whether it truly needs to take priority today. This approach also reduces the chance of people feeling dismissed, as you're making them part of the decision.

For interpersonal conflict between two team members, don't jump in to mediate right away. Instead, ask each person to give direct feedback to the other and try to find a way to work together. Stepping in too quickly can prevent them from building the skills they need to navigate conflict on their own.

A helpful lens I use to avoid this trap is to categorize problems into three types:

Problems that are clearly your responsibility

Things like team-wide blockers, technical alignment, or cross-functional miscommunication.

Problems that are clearly their responsibility

> For example, completing a task, owning a feature, or resolving a peer conflict (with your support, not necessarily your intervention).

Gray-area problems

> Those that could be yours or theirs, depending on the context. In these cases, consider who is best positioned to take ownership based on workload, goals, or development opportunities.

This framing helps you slow down before jumping in and makes it easier to coach your team toward ownership instead of taking the reins by default.

You're part of a team, not a solo problem solver. Set clear boundaries, let others own their responsibilities, and focus on guiding rather than fixing everything yourself. This empowers your team, prevents burnout, and keeps you from being overwhelmed.

AVOIDING MICROMANAGING

When I first became a tech lead, I believed micromanagement was the only way to ensure things were done perfectly. I had a to-do list for every person, verified everything they did, and struggled to delegate. I felt constantly angry, frustrated, and ultimately, burned out.

I learned the hard way that micromanagement isn't sustainable. It limits both the leader and the team, leading to frustration, burnout, and a lack of trust. Micromanaging might feel like control, but it's actually driven by fear: fear that things will go wrong and a lack of trust in both yourself and others.

The turning point came when I learned to let go. I started by reflecting on my actions, getting feedback, and practicing delegation every day. By doing so, I shifted from trying to control everything to focusing on people and their growth.

If you recognize yourself in this, it's time to start letting go. Here's how:

Start with small tasks

> Delegate small, noncritical tasks first, and trust your team to handle them. Gradually increase responsibility as trust grows. More on this in Chapter 6.

Give clear expectations, not step-by-step instructions

> Focus on the "what," "why," and "when" of tasks rather than the "how." Let your team figure out the most effective way to accomplish tasks. Allow yourself to be surprised by their approach.

Cocreate a tracking system

Instead of constantly checking in, ask your team to develop a visible tracking process for their progress. Schedule regular check-ins where they report back. This keeps you informed without hovering.

Practice saying "I don't know"

Start in a one-on-one with someone you trust, then try it in a team meeting. Notice how it feels and how others respond. It helps build comfort with uncertainty and sets the tone that it's OK not to have all the answers.

Admit a mistake

Pick a comfortable setting, such as a one-on-one, to share a mistake you've made. It's a small but powerful step toward letting go of control.

Reflect on your control tendencies

Take time to think about how your need for control impacts your well-being and the people around you. Working with a coach or mentor can offer valuable outside perspective.

Ask for help

This requires an extra dose of vulnerability. Let your team in on your journey by admitting, "I've realized I tend to control too much, and I want to work on it. Where do you think I should start?"

When you focus on guiding rather than controlling, your role as a leader becomes about empowering others. This shift benefits your team and your own well-being. You'll have more mental space, less frustration, and greater trust in your team's capabilities.

Micromanagement often comes up as a reason people want to leave their jobs. No one likes to be watched all the time, living with the constant fear of making a mistake and having no autonomy in making their own decisions. As the saying goes, "People leave managers, not jobs."

If you want your team to thrive and stay engaged, invest in building trust. Let them own their tasks, and allow for mistakes, because that's how people learn and grow (e.g., some of the best learnings come from postmortems). You might be surprised at how capable and motivated your team becomes when they feel genuinely trusted and empowered.

Moving away from micromanagement doesn't mean losing control of outcomes; it's about gaining trust and respect. You'll see results not by doing everything yourself but by leveraging the full potential of your team. It's a shift from "I must do everything to make sure it's perfect" to "I trust my team to succeed, and together we'll grow.

BALANCING BUILDING AND LEADING

Another challenge for tech leads is deciding how much time to spend coding versus overseeing technical decisions.

When I first became a tech lead, I found myself naturally drawn to high-level strategy and discussions, leaving less time for hands-on coding. I enjoyed meetings, brainstorming solutions, and setting direction for the team, but I soon realized I was losing touch with the actual technical work.

At the same time, I've also met tech leads who remain deeply immersed in coding, losing sight of the bigger picture because transitioning to a leadership role feels unfamiliar and uncomfortable. Both approaches come with their own set of challenges.

Determining where you fall on the spectrum between technical and organizational focus starts with honest self-reflection. Imagine a day where you could choose to do only one thing. Would you rather spend it immersed in code, building and debugging, or would you lean toward conversations, planning, and driving strategic direction for your team? Your answers to these questions can reveal where your strengths and interests naturally lie, helping you align your focus as a tech lead.

Recognizing your natural tendencies is important, but to be a successful tech lead, you need to have the ability to switch gears between hands-on coding and strategic oversight based on what your team and project need most.

If you're too focused on coding, it might show up in a few ways. Maybe you're the first to jump on every pull request or technical challenge, leaving little room for others to step in. You might know the codebase inside out, while the rest of the team feels left out of key decisions. If you're skipping meetings or high-level planning, chances are the team is missing the clear direction they need to move forward.

On the flip side, being too focused on oversight can leave you feeling disconnected from the code. You might find yourself zoning out in standups or struggling to understand the technical details the team is discussing. This can frustrate your team when they are discussing technical problems and you're either unavailable or too out of touch to step in.

The key to being an effective tech lead is adaptability. You'll need to adjust your focus based on the project phase and your team's needs. For example:

- When deadlines are tight, you might need to dive into the code and work alongside your team.
- When there's a new feature to plan, you might spend more time away from coding, working with stakeholders and aligning the team on the strategy.

Here are some practical tips for finding balance:

Block time for both

Schedule coding and pair programming blocks into your calendar while keeping time open for meetings, reviews, and strategy sessions. This ensures you're hands-on without losing sight of the bigger picture.

Delegate but stay informed

You don't need to be the one writing every line of code, but stay involved enough to guide technical discussions and ensure the code quality is up to standards.

Engage in code reviews

Even if you can't contribute code regularly, reviewing pull requests (PRs) is a great way to stay connected to the technical aspects without being fully hands-on. I also know tech leads who do PR reviews every morning with the team, all together, not as a line-by-line review but more as a high-level walkthrough, scrolling through the code to share insights and context. If you work with trunk-based development and have no PRs, block core pair programming hours on your calendar.

Ask for feedback

Check with your team and your stakeholders to see if they feel you're striking the right balance. They'll let you know if they need more of your technical input or if they feel you're too involved.

The key to overcoming this challenge is to remain flexible. Some phases of a project will demand more hands-on coding, while others require you to step back and focus on guiding your team. The goal isn't to be perfect in both areas but to adapt to what the situation calls for and ensure you're meeting both the technical and leadership needs of your team.

Key Takeaway

Stepping into the tech lead role can feel overwhelming, but it becomes much more manageable when you approach it intentionally. This chapter was about giving you a structure to grow in a way that fits you. From clarifying what's expected to figuring out where you're starting from, it's all about creating a path you can actually follow.

You don't need to get it all right immediately. What matters is that you keep learning, adjusting, and showing up with purpose.

3

Building Relationships

Everyone in tech understands the value of networking when it comes to advancing their career. It's how you hear about job openings before they're public, skip the application line with referrals, or even land promotions by building trust with managers. Networking gets you insider knowledge about companies, helps you decide if they're the right fit, and gives you a direct line to opportunities that might otherwise take years to stumble upon. The right relationships can fast-track your career and save you from a lot of wasted time and effort.

As a tech lead, the stakes are even higher. Building strong relationships isn't just about you anymore; it directly impacts your team and the teams you collaborate with. The benefits multiply significantly, from creating smoother workflows to opening doors to new opportunities for your team. Yet, I've noticed many tech leads roll their eyes when I bring this up. They see building relationships as time wasted, a distraction from coding, and often fail to connect the dots between stronger relationships and fewer headaches for themselves and their team.

This chapter is here to change the perception that building relationships is a waste of time. I'll show you the value of investing in the right relationships as a tech lead and how these connections can unlock opportunities, prevent unnecessary problems, and make life easier for everyone involved. I'll dive into the relationships you need to focus on, strategies for building and maintaining them, and ways to overcome common struggles like breaking the ice with someone new, keeping relationships strong over time, and connecting with nontechnical people in other departments.

The Value of Building Strong Relationships

One of the teams I led stood out as one of the most successful I've worked with, not just in terms of results but in how we worked together. We delivered reliably, collaborated efficiently across teams, handled conflict in healthy ways, and brought in significant business value. People were happy and proud to be part of it. I'm convinced that one of the biggest reasons for this team's success was the relationships we built, with each other, with stakeholders, and with clients. These relationships didn't just happen on their own. As a tech lead, I made them a deliberate focus. I didn't wait and hope connections would form naturally; I put consistent effort into building and nurturing them every step of the way.

Within the team, we created a space where tough conversations could happen without hesitation. When something came up, we addressed it head-on, tackling issues quickly instead of letting them grow into bigger problems. We challenged each other constructively, not to prove someone wrong but to help everyone grow. Feedback became a daily improvement tool for us. Decision making wasn't just up to me; we debated ideas openly, came to agreements together, and took collective ownership. It wasn't always easy, but it set us apart as a cohesive and high-performing team. If you're curious about how we created this environment, the section "How to Create Psychological Safety on Your Team" on page 188 dives deeper into the specific practices we used.

Building strong relationships with stakeholders, especially on the client side, gave my team a huge advantage. I invested in these relationships consistently, showing up for them in ways that mattered: whether it was being transparent about potential issues, helping them prepare for tough conversations, or stepping up to support initiatives outside my immediate responsibilities, like writing stories or contributing to their ideas.

A great example is advocating for addressing tech debt. Many tech leads struggle to convince stakeholders to prioritize it because it's often perceived as wasted time on the business side. But thanks to the trust I built with our product manager, he didn't need convincing. He trusted our judgment because of the foundation we had built, and when I asked for time to clean up a feature toggle to avoid confusing code that might slow down future reviews or increase the risk of bugs, or address technical debt more broadly, he didn't hesitate to say yes, even if he didn't fully understand the technical details. That trust came from consistent effort: recurring one-on-ones, full transparency around progress and potential risks, and staying responsive even when things got tough.

Here is another example of how building a strong network and staying visible as a team can pay off: when an exciting opportunity came up, like leading a high-profile product launch, we were the first team considered and offered the chance to develop the project because we had consistently stayed visible and well-connected.

When working with other teams, building relationships allowed us to share knowledge and collaborate more effectively. Knowing what they were working on helped us avoid duplicating effort and saved us a lot of time. At the same time, we made sure our work was visible by offering solutions we'd already implemented to others, which built our reputation across the company.

Every time I share these stories and the benefits of strong relationships with tech leads, one question comes up again and again: "If I do all of these, when do I do my job?"

Building relationships isn't something separate from your job.
It is your job.

The good news is, you don't have to do it all on your own. One of the best ways to build a connected, resilient team is to involve others. Delegate. Empower your team to represent and support one another. For example, you might have a staff engineer lead guild meetings or ask a team member to attend a cross-functional sync.

In my case, I encouraged my team to build their own bridges by visiting clients, collaborating with peers in other teams, and being proactive in communication. We set up open Slack channels, made ourselves available, and made clarity a priority. Even if it felt repetitive, we over-communicated to ensure everyone stayed aligned.

Strong relationships don't just depend on you; they're a shared effort. And yes, this work takes time and energy. But the payoff is real: fewer misunderstandings, smoother collaborations, a team that's more motivated and empowered to deliver their best work (Figure 3-1).

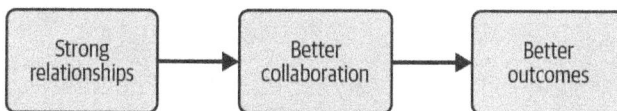

Figure 3-1. *Why building relationships is key*

As a tech lead, you're responsible for developing an environment where people feel safe to speak up, collaborate openly, and trust each other. You may not always be able to predict exactly how or when a strong relationship will help, but it will. The effects ripple through your team's work in ways that matter. It's an investment you and your team make together, and one that always pays off.

How to Build Strong Relationships

At the core of any meaningful connection is trust, and trust is built through four foundational practices:

Effective communication
> Making sure your message lands as intended and that you're also hearing others clearly

Consistent communication
> Staying in touch regularly

Transparent communication
> Sharing openly, including uncertainties, trade-offs, and the "why" behind decisions

Follow-through
> Doing what you say you'll do and showing you can be counted on

Each of these practices reinforces the others. When you communicate clearly, frequently, and honestly, and back it up with action, trust builds naturally. In this section, I'll break these down one by one.

But before we dive into these practices, I want to call out one thing that I believe is a game-changer when it comes to building trust quickly: meeting people face-to-face.

There's something about human connection that no tool or strategy can fully replicate. Face-to-face interaction speeds up trust in a way that's hard to explain but easy to feel. That doesn't mean I'm advocating for full-time in-person work; I've been working remotely for years. But if you have the chance to meet someone in person, take it.

It could be a team lunch on a day when everyone's in the office, a coffee with a stakeholder, an offsite that brings people together, or even a celebration for a big release. These moments of shared presence can set a foundation that makes every future interaction smoother, more human, and more effective, even once you're back to Zoom calls and Slack threads.

This being said, I do know that these in-person moments aren't always possible, and that's OK. What matters most is how you show up in every interaction.

Whether remote or in person, the following practices will help you build strong, lasting relationships grounded in trust.

COMMUNICATE EFFECTIVELY

Effective communication is the foundation beneath everything else. You can be consistent and transparent, but if your message doesn't land, or you misunderstand others, trust starts to erode.

And in order to really understand, you need to really listen. Here are some strategies to develop your active listening skill:

Focus fully on the speaker

Give them your full attention. Silence notifications, minimize distractions, and resist the urge to plan your response while they're speaking. This simple shift shows respect and reinforces that you're present.

Get comfortable with silence

Pause before responding; counting to some fixed number can help (even briefly holding your comment) and give space for reflection or additional thoughts from the speaker.

Reflect to confirm your understanding.

Use phrases like "Did I get that right?" or "Let me try to rephrase what I heard..." to ensure clarity and avoid misunderstandings.

Use open-ended questions

Encourage deeper responses with prompts like "Please tell me more about that..." or "What's your perspective on...?"

Reflect after conversations

Post-discussion, ask yourself: How well did I listen? Did my mind wander? Did I interrupt? This helps build self-awareness over time.

Seek feedback on your listening

Ask team members directly, "What's one thing I could do to make you feel more heard?" This input is invaluable for growth.

So aim to listen not just for information but for understanding. Then, when it's your turn to share, follow these guidelines:

Be clear

Use simple, direct language.

Be structured

Bullet-point key takeaways. Add summaries when things get long.

Be considerate

Tailor your message to the audience: what do they need to know?

Be helpful

Link to docs, add examples, or suggest next steps.

Be reflective

Take a second before you speak or write. Don't just dump your thoughts as they come.

AI assistants can be surprisingly effective at helping you with this, especially because they provide a low-pressure environment for practice, feedback, and iteration. Here's how they can support you:

Practice difficult conversations

Simulate one-on-one conversations or challenging scenarios, like giving feedback, handling disagreement, or delivering bad news. Prompt: "Act as a team member who's unhappy with a project decision. Help me practice how to respond in a calm and clear way."

Improve written communication

Paste your Slack message, status update, or email and ask the AI to check for tone, clarity, and simplicity. You can even request a rewrite tailored to your audience (e.g., nontechnical stakeholders, busy execs). Prompt: "Rewrite this status update to make it more concise and less technical."

Ask for structure and models

AI can help you organize complex ideas into clear, easy-to-understand formats, like Problem–Solution–Impact or Before–After–Benefit. Try prompting "I need to explain a complex idea to my manager. Can you help me structure it in a way that's easy to understand?"

Get feedback on tone and intent

AI can help you assess whether your message might come across as harsh, vague, or passive. This is especially useful when you're unsure how it might land. Prompt: "Here's what I wrote; does it sound defensive?"

Learn by doing

Ask AI to analyze communication examples and suggest improvements so you can learn patterns over time. Prompt: "What makes this message confusing? How would you improve it?"

Communicating well is a skill, and like any skill, it can be learned and refined over time. The better you get at this, the easier it becomes to build trust, influence outcomes, and keep your team aligned, especially in high-pressure moments.

Tip

If you want to go deeper into communication strategies that help build strong relationships, I definitely recommend Nonviolent Communication, a system developed by Marshall B. Rosenberg. You can learn more through his book with the same title, published by PuddleDancer Press, and there are other resources available through the Center for Nonviolent Communication (*https://www.cnvc.org*).

COMMUNICATE CONSTANTLY

Constant communication is about staying connected and ensuring a steady flow of information between you, your team, and stakeholders. It's not just about being available; it's about building a culture where sharing, listening, and collaboration become a regular part of your work.

Here's how to make this work in your role:

Hold regular check-ins

One of the most reliable tools for maintaining open lines of communication is recurring one-on-ones. These aren't limited to just your team members. I try to schedule regular check-ins with anyone I collaborate with, whether it's my manager, a stakeholder, or a peer. I'll take the opportunity with anyone who's open to a regular conversation, especially early on, because it's a great way to build rapport and trust.

These one-on-ones serve multiple purposes. They help build trust, catch issues early, and make sure alignment doesn't drift. They also create space for deeper, more honest conversations than what you'll get in a standup or async chat. Think of them as part of your communication hygiene; if you keep them up regularly, everything else tends to flow more smoothly.

You might be thinking, "But this will take up all my working time." And yes, it does take time. But this is a big part of your work. Investing early in relationships saves you hours later. When these relationships are already in place, everything from resolving conflicts to asking for support becomes easier. Of course, you don't need to spend your entire week in one-on-ones. The frequency and format should flex based on the context. Your team may need more consistent support, while stakeholders may change or have less availability.

The key is to make communication a habit, not just a reaction to problems. Build those bridges early and keep them open.

Communicate progress continuously

Keep everyone, both your team and your stakeholders, in the loop throughout the journey, not just when you've reached a milestone or solved a major problem. Share updates on progress, blockers, mistakes, and changes in direction, about both the good and the uncomfortable.

Have a clear, shared way to track progress that's visible and easy to access. Whether it's a roadmap, dashboard, tracking board, or shared doc, it should reflect what's happening and help everyone stay aligned. Even more important than the tool is the habit: make time to review progress together, through weekly check-ins, async updates, or retrospectives.

Avoid being a knowledge silo

If you're the only one who knows something, it's a problem. You don't want to become a blocker, or worse, the single point of failure. When people rely only on you for a particular topic or piece of information, it creates bottlenecks and slows down the team. It also signals that you may not trust others to take ownership or contribute.

Instead, make it a habit to share information proactively with both your team and your stakeholders. Don't keep things to yourself. Whether it's context from a leadership meeting, technical decisions, lessons learned, or simply something you're noticing, bring it into the open. The more you share, the more others can participate, make informed decisions, and act independently.

By passing along insights, even small ones, you're showing trust and earning it in return.

Staying connected through constant communication is how trust starts to grow. Showing up regularly, keeping people in the loop, and sharing information openly builds a solid foundation for your relationships.

BE TRANSPARENT IN YOUR COMMUNICATION

Transparency can be complicated for tech leads. Deciding what to share, when, and how much depends heavily on the situation and your team's context.

Many tech leads, myself included, fall into the habit of keeping things to themselves. The reasoning often stems from trying to protect the team, not knowing how to frame the information, or simply wanting to maintain control of the narrative. This can extend to stakeholders, where fears of repercussions or "they don't need to know yet" lead to delays in sharing issues, hoping they'll resolve before anyone notices. But the problem is that information flows. When people discover something from someone else, the situation often feels worse than if it had come directly from you. Transparency builds trust; trying to hide things undermines it.

Breaking out of the habit of withholding information isn't easy, but you can work on it gradually, one step at a time. Here are practical strategies to help you integrate transparency into your daily routine:

Be honest about both progress and problems

Transparent communication means that if something prevents progress, you let your team and stakeholders know as early as possible so they can plan accordingly. If something's worrying you, whether it's a risk, a misalignment, or something that just feels off, raise it. Don't wait until it becomes a bigger issue. Addressing things early, together, gives others the chance to contribute and often leads to better decisions.

For example, let's say your team underestimated the complexity of integrating with another system, and what initially looked like a small task is now causing delays. Or maybe a risk you flagged earlier has materialized into a real blocker. The moment you realize this will impact timelines, communicate it, even if you don't have all the answers yet. A simple message to stakeholders like "We hit an unexpected issue with the integration that might affect the release date. We're working on a mitigation plan and will update you tomorrow" is far better than delivering bad news late or having people find out from other sources that things aren't going according to plan. Early updates give people the chance to adjust.

Provide balanced feedback

Another way to build trust is by giving both positive and improvement feedback. It takes courage to be open about areas for growth, but this balanced approach shows genuine investment in others' development.

Offering feedback this way helps people know you're honest with them and truly interested in their growth, not just in keeping things positive. It's the combination of encouragement and constructive insight that reinforces trust and strengthens your relationship.

Lead with vulnerability

A lot of tech leads go out of their way to hide their mistakes, fearing it might make them seem weak or cause them to lose their team's respect. The truth is, admitting when you don't have all the answers or when you've made a mistake often has the opposite effect: it makes you relatable and authentic. People already know no one is perfect, and acknowledging this builds trust.

For example, if a technical decision you advocated for led to delays or issues, admit it to your team and stakeholders: "This approach didn't work as I hoped, and it's caused these delays. Here's what we've learned, and this is how we're addressing it." This kind of openness builds stronger connections, creating a culture where your team feels safe to own their mistakes too.

If you find yourself holding back, these strategies can push you out of your comfort zone and help you confront the fear of sharing. Over time, you'll develop new ways to handle interactions and manage challenges directly, instead of staying in the "safe zone" and risking bigger issues. Reflect on your reasons for withholding information. If it's simply "I'm unsure how my team will react," share it anyway. The real challenge is learning to navigate their reactions effectively.

On the flip side, there's oversharing. Let's say your engineering manager shares news of a potential company restructuring, and you're one of the first to know. You don't yet have concrete details, and a big product launch is two weeks away. If you share this prematurely, you risk creating unnecessary anxiety and distraction, especially since you can't provide clarity or answer questions yet. Sharing too much too soon can harm focus and productivity.

There's also a category of information that should never be shared more widely, whether out of respect, to maintain trust, or for legal reasons. This includes personal circumstances, details about layoffs, performance issues, confidential feedback shared in one-on-ones, or financial information that not everyone has access to. In extreme cases, mishandling this kind of information can even result in legal consequences.

So how do you find the balance? Transparency is a judgment call that depends on context. The "right" amount of sharing varies based on timing, audience, and the nature of the information. With time you will learn to read the room and adapt your communication accordingly.

Start by asking yourself a few key questions:

Is it OK to share?
> Check with your EM if this is something that can be shared with the team. If not, the decision is clear. If it's up to you, ask for their perspective to help guide your choice.

What additional context do I need?
> Gather more details about the timeline and potential impacts from your EM or other tech leads. More information helps you decide what's worth sharing.

What's the team's maturity level?
> Teams with more junior members may react more emotionally to uncertainty, while senior teams might handle ambiguity better. Tailor your approach based on their capacity to process and stay focused.

Neither extreme, keeping information or oversharing, is the answer. Finding the middle ground is about intentional experimentation. Start by recognizing your default tendency. If you lean toward withholding, challenge yourself to share earlier than usual and track what happens. Did the reaction align with your fears, or was it better than expected? Gather feedback and adjust based on the results.

For me, my instinct has always been to keep things to myself. It felt safer and gave me a sense of control. But over time, I realized that sharing, even when it made me uncomfortable, almost always led to better outcomes. When I was up front about issues or potential delays, my team and stakeholders felt included and trusted me more.

Transparency helps your team align better, work with less confusion, and feel valued. It creates credibility with stakeholders and strengthens relationships across the board.

FOLLOW THROUGH ON COMMITMENTS

The simplest way to build trust is to say you'll do something...and then do it.

Actions speak louder than words. You can communicate clearly, frequently, and transparently, but if your actions don't align with your words, trust breaks down fast. Follow-through is where trust becomes real.

People need to see that they can count on you. That means doing what you say you'll do, whether it's shipping a feature, checking in on a topic, or raising a difficult issue with leadership.

That might look like agreeing to review a team member's design doc by Friday and actually carving out time by Thursday to leave thoughtful feedback. Or telling your PM that you'll talk to the infrastructure team and then closing the loop a few days later without needing to be chased for updates.

Even if the task isn't fully done yet, don't wait to communicate. If a team member asks a question in Slack and you said you'd follow up, do it. Even something as simple as "Still looking into this; I haven't forgotten" can go a long way in reinforcing trust.

Things won't always go to plan. If something slips, say so early. If you underestimated a task, let people know. A quick message like "I'm still working on this; running behind but I'll share a draft tomorrow," shows accountability and keeps others from worrying or guessing. And if you drop the ball on a follow-up, acknowledge it. Let people know where things stand and what's next. Silence only creates confusion.

Follow-through also matters when things are ambiguous or uncomfortable. A team member once asked me for a raise. Being in a startup, we didn't have a formal process, and I knew budget constraints meant it was unlikely to be approved. Still, I promised to raise the topic with the CEO, and I did. The outcome wasn't a yes, but we agreed to revisit the request in six months, which gave the team member clarity and a timeline. Too often, I see people left in limbo on requests like these because it's not an easy conversation. But avoiding hard conversations doesn't make them go away. People should at least know where they stand.

Reliability is built through repetition. It's about showing up consistently, keeping your word, and owning your actions, especially when it's hard. Over time, people start to notice: when you say you'll do something, it gets done. And that kind of reputation becomes one of the most powerful leadership tools you can earn.

Who to Build Relationships With

As a tech lead, your work doesn't happen in a vacuum. You're not just collaborating with your immediate team but also navigating relationships with your manager, cross-functional peers, and a range of stakeholders across the organization. Each of these relationships plays a different role in your success and requires a different approach. In this section, I'll show you how to apply the principles of building strong relationships from the previous section to strengthen these connections.

TEAM MEMBERS

These are the people who shape your daily experience. The strength of your bond with your team members directly impacts not only how well you work together but also the quality of your outcomes. Building these relationships requires deliberate effort and consistency beyond occasional conversations.

Here are actionable strategies to build strong relationships within your team:

Set up one-on-ones

One-on-ones are especially important with your team.

One of my first moves when joining a new team is to set up recurring one-on-ones with each team member. The frequency might adjust, or we might skip one occasionally, but the important part is that they're in the calendar. Just having that space booked gives your team confidence that you're there, that you're invested, and that you'll continue to show up.

These meetings are so much more than status updates. In fact, updates should be just a small part of them. One-on-ones are your best chance to notice when someone's struggling, whether with workload, motivation, or something personal. They create a space where people feel safe to be honest. If someone is underperforming or going through a tough patch, this is the time to dig deeper, offer support, and align on a way forward.

If you want to dive deeper into how to approach these conversations with your team, check out Chapter 4. It covers everything from setting them up to dealing with common challenges, like when someone shuts down or you're unsure how to help.

Make feedback routine

Treat feedback as a natural and regular part of daily interactions rather than a formal event. Consistently sharing both positive and constructive feedback builds trust and shows your genuine commitment to your team's development. Addressing areas for improvement may require courage, but combining encouragement with actionable insights makes team members feel supported and motivated to improve. Encouraging your team to exchange feedback among themselves amplifies this effect, strengthening connections and collaboration across the group.

If you want to go deeper into how to ask for, give, and create a healthy feedback culture in your team, check out Chapter 5. It covers everything from the five principles of good feedback to the practicalities of making it work day-to-day, including the most common challenges tech leads face and how to overcome them.

Say "I don't know" more often

A simple and effective way to connect with your team and build trust is to show that you don't have all the answers, and that's OK. Saying something as simple as "I don't know" can be a game-changer. It removes the pressure to pretend, accelerates your learning because you're more likely to seek answers, and creates a safer space for others to admit when they're unsure too. Using phrases like "I don't know, but I'll find out" or "Let's figure it out together" not only shows humility but also strengthens your credibility.

Give your team more autonomy

Fear of losing control is common for tech leads, often leading them to micromanage: tracking everyone's progress, constantly checking in, or reacting negatively when things don't go as planned. This behavior creates discomfort and causes team members to avoid you, harming your relationship with them.

Strike a balance between being available and overbearing. Be approachable for questions and problems, but also give your team the space to work independently, make mistakes, and learn without fear of your reaction. The right amount of independence and guidance isn't one-size-fits-all. More junior team members might need more structure and might reach out more often, while senior engineers often thrive with greater flexibility. Tailor your level of involvement based on the individual or team's experience and confidence.

A powerful way to build autonomy is through delegation. Identify a task (you can get ideas in Chapter 6), select the right person, and set clear expectations. Offer support but then step aside. This shows your team you trust their skills: *"I trust you to take care of this."* When team members feel trusted, they're more likely to trust you, strengthening your relationship. Plus, prioritizing their growth shows you value their development, which they'll deeply appreciate, building even stronger connections.

Focus on solutions, not blame

When things go wrong, don't make it about pointing fingers; focus on fixing the problem. A team that knows they won't be blamed for mistakes will feel safer to share challenges and work together to solve them. Mistakes are inevitable, but how you handle them can either strengthen or break your team.

Take a production incident as an example. If your response is *"Whose fault is this?"* you're setting a tone of fear and mistrust. People will start hiding their mistakes, avoiding accountability, and holding back potential issues because they're afraid of how you'll react. But if you shift the focus to "How can we solve this?" or "What ideas do you have to fix this?" you send a different message: mistakes happen, and that's OK, as long as we learn from them and tackle them together.

Handling things this way creates opportunities for your team to bond. There's no better team-building exercise than solving a tough problem together. By consistently approaching issues with a solutions mindset, you'll build a culture of shared responsibility, where the whole team feels accountable for both successes and failures. Over time, your team will follow your lead and stop looking for someone to blame, focusing instead on how to move forward together.

Note

How you lead when things go wrong says a lot about you as a tech lead. Those moments show your true character, and they're where trust is built, or lost. Choose wisely.

These strategies, when applied consistently, will strengthen the connections within your team, creating an environment built on trust, collaboration, and continuous growth. With these relationships in place, your team will feel supported, empowered, and equipped to deliver better results.

MANAGERS

Your first team stakeholders are your managers, and they can either help you or be a real roadblock; it usually depends on your relationship with them, not just on having good arguments. When you've already established a connection, your suggestions and decisions tend to land better:

Make one-on-ones a priority

If your manager or stakeholders don't initiate one-on-ones, take the lead and set them up yourself. These meetings don't need to be as frequent as the ones with your team; every two weeks or even monthly can work, depending on the stakeholder. The important part is consistency.

Show up prepared and organized to make their job easier. Bring notes, ask thoughtful questions, and show genuine interest in their work. Often, tech leads hesitate to request these meetings, assuming stakeholders have more important things to do, but the truth is they also benefit from these conversations. One-on-ones are an opportunity to address potential issues early, align on priorities, and gain valuable insight into broader organizational goals. If they're too busy, they can decline. All you have to do is ask.

Highlight your team's work

For all the effort your team puts into their work, you should match that by finding ways to make it visible to stakeholders.

A great starting point is investing in showcases. These are presentations where your team shares what they've been working on, like new features, progress, or key learnings, with others outside the team, especially stakeholders. They're a chance to keep everyone in the loop and show the impact of your collective work. Treat them like value-delivering tasks, not just routine meetings. Add them to your team board, set time aside to define what you want to highlight, prepare the presentation thoroughly, and rehearse your delivery. When done right, showcases can bring tremendous visibility to your team's work and trust from stakeholders. For example, my team's well-prepared showcases made us the first choice when a high-profile product opportunity appeared.

The key to building trust during these showcases is transparency. Don't just celebrate wins but also share the challenges your team faced and the lessons learned. This openness reassures stakeholders that they can trust you to deliver while being honest about what didn't work. Ending with a feedback session is a cherry on top, encouraging conversation and engaging stakeholders who might not otherwise get involved.

This process doesn't rest solely on you. Rotate team members to lead showcases; it gives them visibility, strengthens their skills, and boosts their pride in the work.

Outside of showcases, one-on-ones with stakeholders can also be an opportunity to share team successes, as well as to build personal connections. Even something as simple as sharing progress updates in public Slack channels can make a difference; you never know who might notice.

Ask for mentorship

One of the easiest ways to build new relationships, especially with people you don't interact with daily, is to ask for mentorship. It doesn't have to be formal or long term. A simple request like "I've noticed how well you handle stakeholder conversations. I'd love to learn more about your approach. Would you be open to a quick chat sometime?" can open the door to a deeper connection.

People generally appreciate being asked for advice or perspective, especially when it's specific and genuine. It shows that you respect their experience and want to learn from them, and that builds trust fast.

Sometimes, what starts as a one-off chat grows into a longer-term mentorship relationship. You may find yourself checking in regularly, exchanging updates, or turning to them during tough moments, without ever needing to label it as formal mentorship. Just like any strong relationship, it develops naturally over time.

Managers are a great starting point, even if they're not in your direct reporting line. They often bring a broader organizational view and can help you see how your role fits into the bigger picture. But don't overlook experienced ICs either; they can share incredibly useful tips on how to lead without official authority or how to handle tricky team dynamics.

The key is the conversation. Reaching out like this creates new allies across the organization and builds relationships that may benefit you (and them) in unexpected ways later on.

Show up at events and use casual moments to build connections

All-hands meetings, leadership events, or casual gatherings like coffee breaks and holiday parties are perfect chances to connect with stakeholders and managers. Even if you find these events boring, they're a great way to build relationships. A quick comment like "I really enjoyed your talk on [topic]" can leave a positive impression and make follow-ups easier.

If a manager offers an informal coffee chat or an "Ask me anything" session, jump on it. Showing up, even when others shy away, sets you apart and opens doors for more meaningful conversations.

By investing in these relationships, you'll find that managers are more likely to support your initiatives, making things smoother for both you and your team.

OTHER STAKEHOLDERS

Your team doesn't work in isolation; every move you make will impact other teams and roles: product managers, designers, cross-functional departments like sales, finance, legal, customer service, and, of course, your users or clients (if applicable). These stakeholders also influence your work and can directly affect how effectively your team achieves its goals.

Here are a few ways to make these connections work for you and your team:

Open cross-team communication channels

Let's take a common example: your team depends on another team to deliver part of a project. Their progress directly impacts your timeline, and vice versa.

The easiest way to ensure smooth communication is by setting up a dedicated Slack channel or space for both teams. This becomes a central hub for updates, questions, and discussions.

Recurring check-ins are another great way to keep everyone aligned. These don't need to be long or formal, just a quick sync to share progress and address potential blockers.

Of course, you don't have to manage this relationship alone. You can pair up with someone from your team who has an interest in the topic or assign them to take ownership of the communication.

This not only keeps updates flowing smoothly but also gives your team a chance to build relationships with the other team, a win for everyone involved.

Ask for help, and offer it

Asking for help is one of the easiest and most underused strategies to build relationships. Instead of wasting days stuck on a problem, tap into the collective knowledge of your organization. Post questions in general or team-specific Slack channels like "Has anyone worked on this before?" or

"Can someone help with this?" You'd be surprised how often people are willing to assist.

Similarly, take the initiative to help others. Answer unanswered questions in Slack or reach out to teams facing challenges you've already solved.

Build cross-team collaboration spaces

Host monthly tech lead chats: set up recurring sessions where tech leads from different teams come together to share updates, troubleshoot challenges, and brainstorm ideas. To make the most of these sessions, consider assigning a rotating facilitator and preparing a lightweight agenda in advance. This helps keep the conversation focused while still leaving room for organic discussion and peer learning.

Create company guilds: establish groups around topics like architecture, frontend development, or cloud migration. Open these guilds to anyone interested, whether engineers, architects, or other stakeholders, to encourage knowledge sharing, alignment, and innovation across teams.

Virtual coffee chats (often facilitated by Slack apps like Donut, Alfy, or similar) are an underrated way to connect with colleagues across your organization. While many companies offer these, few people take full advantage of the opportunity. Yet, I've never met a developer who participated and regretted it. These informal chats can spark unexpected connections, open doors to collaboration, and help you build a network that might come in handy when you least expect it. You never know when the person you chat with could be the key to solving a future challenge or opportunity.

Support other teams' initiatives: encourage your team to participate in hackathons, brainstorming sessions, or workshops led by other teams. You can also show up to their demos, give feedback on deliverables, comment on RFCs, or contribute ideas in shared documents and discussions.

Invite diverse perspectives to brainstorming

For product kick-offs or major planning sessions, bring in voices from other departments like customer service, sales, or marketing. Instead of only the PM talking to them separately, I've found it more effective to bring everyone together for their take on potential product challenges and solutions up front. Their perspectives often surface challenges and opportunities the core team might miss. Just ask! They can always say no, but the payoff can be significant when they say yes.

These strategies can help you start building connections with your team and different types of stakeholders.

Common Challenges and How to Overcome Them

Building relationships comes with its fair share of challenges, especially as a tech lead. The three most common ones I see tech leads face are breaking the ice with new connections, maintaining strong relationships over time, and effectively bridging the gap with nontechnical stakeholders.

For some, these conversations feel easy and natural. They enjoy small talk and are quick to build rapport. But for many others, starting and sustaining these kinds of interactions can feel awkward, mentally draining, or just unclear. Whatever the reason, know that you're not alone and that this is something you can absolutely get better at through practice.

In this section, I'll dive into the three most common relationship-building challenges mentioned previously and share actionable strategies to help you navigate and overcome them.

IT FEELS AWKWARD TO START CONVERSATIONS

Breaking the ice can feel daunting, but without that initial moment of connection, relationships can't begin. Many tech leads overthink the first step: "What should I say? How do I start?" Instead of waiting for opportunities to come to you, take the initiative.

A presentation is one of the easiest ways to start a conversation. Speakers expect interaction, so after their talk, show appreciation and highlight a specific point you liked: "I really enjoyed your take on..." Asking a thoughtful question about their content can make the interaction more meaningful, and they'll remember your genuine interest.

If you're working on a shared project, use that as common ground. Reaching out with "I heard you're tackling a similar challenge; how did you approach it?" opens doors naturally.

Similarly, asking for help is an underrated strategy. Most people love sharing their expertise. Be specific: instead of a vague "hello," start with "I heard you're the go-to for this issue. Can you help?" This clarity increases the chances of a response.

Don't underestimate the power of active listening. When someone responds, be fully present: pay attention to their words, tone, and body language. Reflecting back what you heard or asking clarifying questions ("So, do you mean...?") shows that you're not just waiting to speak but are truly engaged. Instead of jumping to your next point, you can use what the other person just said to carry the conversation forward. People are far more likely to remember someone who made them feel heard than someone who had the perfect opening line.

Also, don't miss opportunities to help others. Answer Slack questions no one has addressed, tag someone who might know, or offer your insight. People notice this effort.

Finally, create spaces where others can approach you (one of my favorites). Host showcases, lead meetings, or run Q&A sessions. It's an easy way to start building connections without having to reach out yourself, as you're creating a space where they can approach you.

The more intentional effort you put into breaking the ice, the easier it becomes, and more and more ideas will come up.

HOW DO I KEEP RELATIONSHIPS STRONG OVER TIME?

Building relationships is only the first step; maintaining them requires consistent effort. Over time, people's priorities shift, and so does the context of your interactions. To keep relationships strong and meaningful, you need to stay intentional and adaptable.

Here are actionable strategies for nurturing these connections over time:

Keep track of people

Make it a habit to jot down memorable details from conversations right after, such as personal interests or key points they've shared. Whether it's a note about their favorite hobby, a challenge they mentioned, or a skill they want to improve, referencing these details in future conversations shows that you care and strengthens the connection.

Follow up

When someone helps you, take the time to circle back and share the results. This shows appreciation and keeps the connection alive. Even small updates matter. For example, you could say, "Thanks again for the tip on [topic]. I tried it, and here's what happened: [result]."

If someone asks you a question you couldn't answer right away or you left a conversation open-ended, always follow up. Even if it's just to say, "I'm still working on this" or "I couldn't find an answer," it shows reliability and keeps the dialogue flowing.

Create recurring touchpoints

Consistency is everything when it comes to maintaining relationships. Avoid skipping one-on-ones with your team or stakeholders, even if it feels like there's nothing urgent to discuss. These moments aren't just about updates; they're about building rapport and staying connected.

Make it easy for others by coming prepared: bring updates, questions, or potential issues, and leave space for them to share their own thoughts.

Prioritize and choose battles wisely

Not every relationship requires the same level of attention. For instance, if you're collaborating with another team to deliver a critical feature, it makes sense to focus on building a strong connection with that team and the stakeholders impacted by the feature. Meanwhile, relationships that aren't immediately relevant can temporarily take a backseat. Regularly reassess your priorities as projects and goals shift to ensure your energy is directed where it matters most.

Similarly, not every disagreement is worth pursuing. Focus on common ground rather than being right. For example, agreeing with your product manager to move retrospectives from Friday to Monday, even if it's always been Friday, is less impactful than advocating for prioritizing a crucial technical-debt task that could directly affect your product's quality.

Relationships are like living things: they thrive with care and effort but can endure beyond immediate interactions. Once trust and reliability are established, they become self-sustaining. For example, if you've built a strong connection with a stakeholder during a project and reconnect months later, that trust still holds, and you can pick up where you left off. This was clear to me when I transitioned to being self-employed (a solopreneur): past clients, coworkers, and managers reached out with opportunities years later.

Investing in lasting relationships also contributes to building a foundation for your long-term career: there is a high chance you will run into the same people in the future.

WE JUST DON'T SPEAK THE SAME LANGUAGE

Usually, this means struggling to get on the same page, like you're both talking but not fully understanding each other's perspective. Often, it's because they're deep in business lingo while you're explaining things technically.

Here's how to bridge that gap:

Step into their world

Start by mentioning how important it is for you to figure this out together and why. This will set the ground for conversation, showing them that you are not going for a fight. Learn what they care about. Everyone's viewpoint comes from their role's priorities, so start by understanding what drives their decisions. Instead of trying to prove to the other person you are right, start by deep diving into their reasoning and arguments. One strategy that I use in these cases is to assume the other person has a valid point, even if I don't fully see it yet. I try to understand their reasoning deeply, believing there's something valuable I might be missing.

Ask clarifying questions

Don't assume their intentions or goals. Ask "What's the main priority here?" or "What's the outcome you're aiming for?" Reflect back their answers: "So if I understand correctly, you mean...?" This one simple question can make a huge difference. It may sound small, but repeating back my understanding has helped me avoid countless misunderstandings. Often, the reply is "No, that's not what I meant," revealing that we weren't actually aligned at all. Most problems in tech teams come from everyone assuming they're already on the same page when they're not.

Find shared goals

Once you understand their goals, look for common ground. For example, if they're focused on deadlines and you're focused on code quality, point out how both efforts contribute to a successful product launch. Shared goals help align your efforts.

Use their language

Explain your points in terms that resonate with their priorities. Instead of saying, "We need to refactor," frame it as "This will reduce risks and save time later." Adapting your language shows effort to meet them halfway, making your ideas more relatable.

Building a shared language doesn't happen overnight, especially when you're coming from different domains. But with curiosity, patience, and a genuine effort to understand the other person's perspective, it becomes much easier to bridge the gap. Remember: it's not about winning an argument; it's about building alignment and trust. And that starts with showing that you're willing to meet people where they are.

BUILDING RELATIONSHIPS AS AN INTROVERT

Not everyone finds it easy to strike up conversations or navigate social dynamics, especially if you're more introverted. The expectation that tech leads should be constantly engaging can feel overwhelming when your natural preference is for quiet focus or one-on-one interactions.

But being introverted doesn't mean you can't build strong relationships; it just means your approach may look different. You might skip the big team lunches and instead build rapport during recurring one-on-ones. You might not jump into every Slack thread, but when you do, you add thoughtful insight. You may not talk the most in meetings, but when you speak, people listen.

People often assume I'm an extrovert. I'm not. I don't get energized by frequent small talk or bouncing between group conversations; I need time away from people to recharge. But I show up and do it anyway, because building strong relationships is part of the job, and I've found ways to do it that still work for me. And I'm not alone. Some of the strongest tech leads I know are introverts. They earn trust through consistency, deep listening, and thoughtful questions that help others feel seen.

If you're introverted, here are a few strategies that can help:

Leverage one-on-ones
> These offer a quieter, more focused setting for building trust. You don't have to perform or think on your feet the same way you might in a group setting. Use these conversations to ask meaningful questions, check in on goals or concerns, and share feedback in a way that feels more natural than speaking up in a crowd.

Prepare, prepare, and then prepare some more
> Many introverts feel more confident when they've had time to gather their thoughts. Before meetings or conversations, jot down key points you want to raise, questions you want to ask, or updates to share. This preparation helps you show up more clearly and assertively, even if the setting isn't your comfort zone.

Use async communication to your advantage
> You don't always need to speak up in real time. Writing a thoughtful message in Slack, leaving comments in a document, or summarizing

your thinking in an email can build your visibility and influence. Written communication lets you contribute on your terms without the pressure of having to respond instantly.

Set boundaries and recharge

As an introvert, your energy is finite, especially in high-interaction roles. Block out time for deep work, avoid back-to-back meetings when possible, and take short breaks to reset. When you protect your energy, you'll show up more present and intentional in the interactions that truly matter.

You don't need to change your personality to succeed as a tech lead. You can definitely be a great tech lead as an introvert. You just need to find a rhythm that works for you and start with the relationships that matter most.

Key Takeaway

Building relationships is about creating the conditions where collaboration, trust, and leadership can thrive. You don't have to become someone you're not, and you don't have to get it perfect. What matters is that you show up with intention, curiosity, and consistency.

Not every conversation will land, and not every relationship will click, but over time, your willingness to invest will shape how people experience working with you.

So pick a place to start. Reach out. Ask questions. Listen well. And remember that building relationships is a shared effort. It's a two-way street. Use these strategies to connect, but also recognize that the other person has to meet you halfway. All you can do is give it your best effort.

4

Running One-on-Ones with Your Team

One of the very first things I do when stepping into a tech lead role with a new team is set up recurring one-on-ones with each team member. While the collaborative dynamic of the team as a whole is essential, these individual one-on-ones are where you can build personal connections, understand unique perspectives, and address concerns that might not surface in group settings. Before diving into the tech stack, analyzing the codebase, or meeting stakeholders, I prioritize these meetings, because the team's collective success starts with understanding and supporting each individual.

That said, there's an important distinction worth calling out early. In some companies, one-on-ones are formally owned by engineering managers, and that can lead to some confusion about whether tech leads should still run their own. My view is simple: yes, you absolutely should. Even if there's some overlap in topics, that doesn't mean the conversations are redundant. It just means you'll need strong alignment with your EM and clarity with your team about your role in supporting them.

Take growth, for example. An EM might manage the formal side, including things like promotions, compensation processes, and official career frameworks, but that doesn't mean people's growth isn't also your responsibility. Quite the opposite. Their development impacts the team's outcomes, and by extension, your success as a lead. As someone who works closely with them day to day, you're often in a better position than the EM to support their progress in practical, immediate ways. That's why these conversations matter, regardless of reporting lines.

These early conversations set the tone for your working relationships and help you establish trust and rapport from day one.

In this chapter, I'll walk you through how to get the most out of one-on-ones: how to structure them, what to talk about, how to build trust, and how to avoid common pitfalls. Whether you're new to the role or looking to uplevel how you approach these conversations, the ideas in this chapter are designed to help you make every one-on-one count, for both you and your team.

The Value of One-on-Ones

One-on-ones are the most powerful tool you have as a tech lead. They will make your life easier, strengthen your relationships, and support the growth of your team members.

First, they help you stay ahead of potential issues. One-on-ones create space to deliver feedback, identify underlying problems, and even find tasks to delegate. They give you a better understanding of what's really going on in your team and help you tackle challenges before they escalate.

Second, they strengthen relationships. These conversations allow you to connect with your team members individually, creating a space for trust and open dialogue. The questions they ask and the feedback they share offer invaluable insights into their interests, concerns, and goals.

Finally, they drive growth. One-on-ones are where you provide the guidance and support your team members need to take on new challenges and develop their skills. They're a direct investment in the team's success, and yours.

MAKE YOUR LIFE EASIER

One-on-ones aren't just for your team members; they're for you, too.

Ever had feedback you wanted to give but kept putting off? A one-on-one is the perfect space to deliver it. It's your chance to stop waiting and address things before they grow into bigger problems (although timely feedback should never be replaced; more on this in the section "How to Give Useful Feedback to Your Team" on page 116).

They're also a window into what's happening behind the scenes. You'll pick up on things that don't surface in team meetings: frustrations, potential conflicts, or underlying issues. For example, during a one-on-one, a team member might express frustration over unclear expectations from another department, something they wouldn't feel comfortable raising in a group setting. This gives you the chance to address the issue privately, clarify responsibilities, and prevent a broader misunderstanding from escalating. Asking the right questions helps you uncover and resolve these issues early.

And let's not forget delegation. Want to get some things off your plate? Use one-on-ones to share what you're working on and see what sparks interest. Ask, "Would you be interested in taking over this initiative?" You'll be surprised how often someone is eager to step up.

One-on-ones also give you a unique view into your team. Pay attention to the questions your team members ask; they're often just as revealing as their answers. For example:

- "How am I doing in this area?" signals they're keen to grow and want your guidance.
- "What's happening with [initiative]?" might mean they're looking for ways to contribute more or feel left out of the loop.

These moments open doors to conversations that help you understand your team better, identify their motivations, and anticipate potential challenges.

One-on-ones are like your leadership cheat code. When you invest time in understanding your team and building strong relationships, you'll find it easier to align on goals, address conflicts, and keep everyone moving forward together.

STRENGTHEN INDIVIDUAL RELATIONSHIPS

In Chapter 3, I talked about how important it is to build strong connections with your team and the impact those relationships have on your team's results. One-on-ones are one of the best tools you have to build those relationships.

These conversations go beyond surface-level work interactions. They create a dedicated space for trust and open dialogue. Showing up consistently for one-on-ones sends a clear message: "I value your time, your input, and you as a person." It's not just about work updates; it's about showing you care, whether it's chatting about their weekend plans or diving into their career goals.

One-on-ones also give you the chance to connect with your team members individually, in a way group settings never can. Talking about personal interests, challenges, or aspirations helps to humanize the relationship and build a stronger bond over time.

And that's why canceling one-on-ones is one of the fastest ways to break trust. If you make a habit of pushing them aside, it sends the message that other things are more important than your team members. Of course, emergencies happen, but canceling without a solid reason, or doing it repeatedly, is a clear recipe to erode the connection you've worked to build.

Commit to these meetings and prioritize them. They are your chance to strengthen relationships, understand your team on a deeper level, and build the foundation for collaboration and trust.

SUPPORT PEOPLE'S GROWTH

The tech lead role varies widely across organizations; sometimes it includes formal line management and other times it's a senior IC position without direct people management responsibilities. Regardless of where your role falls on that spectrum, one thing remains true: supporting the growth of your team members is part of the job. Their development directly influences the team's outcomes, and, by extension, your success as a lead.

One-on-ones are a golden opportunity to support that growth.

Start by emphasizing the value of a growth plan (see the section "Developing a Personal Growth Plan" on page 38). If your team members don't have one, help them define it. Talk about the skills they want to develop, areas they want to improve, and how their growth aligns with the team's goals.

Once you have a clear sense of what growth looks like for each individual, mentoring and coaching become two of the most effective ways to help them get there. Before diving deeper, it's worth addressing a common question: what's the difference between mentoring and coaching, and which one should I be doing as a tech lead? See Table 4-1 for a breakdown.

Table 4-1. 1 Difference between a mentor and a coach

Mentor	Coach
Shares experiences, opens doors, offers guidance and advice.	Helps others discover solutions, supports journeys, challenges their thinking.
Better suited for junior developers: "When I faced a similar issue, I found that breaking it down like this helped. Would you like me to walk you through it?"	Better suited for senior developers who have experience but need help choosing the right approach: "What options have you considered? What do you think might work best and why?"

At their core, mentors and coaches are tools to support growth, but they differ in approach. Mentoring often involves giving advice. People come to you for your expertise and experience; they expect your input. That's perfectly valid. Just remember: your experience might not map directly to theirs, so after you share, help them reflect on how it applies to their situation and what they might take from it.

Coaching, on the other hand, treats advice as a last resort. Instead of offering answers, it leans on asking deeper questions to help the other person uncover their own solutions. In fact, when I find myself in a coaching conversation, I often say, "Let's explore this together, and if by the end you still feel stuck, I'll share my take." Most of the time, they find their own answer; they just need someone to help them navigate toward it.

Tip

For a practical introduction to coaching, *The Coaching Habit* by Michael Bungay Stanier (Page Two) offers a simple framework of powerful questions that any leader can start using right away; no formal coaching background is required.

That said, don't get too hung up on definitions. As a tech lead, you don't need to obsess over whether you're being a coach or a mentor in a given moment. What matters is utilizing the techniques each of them brings to the table to support your team's growth.

Both methods are focused on helping others grow, and one of the first things they offer is accountability.

To provide accountability, start by helping your team members articulate clear, realistic goals. Then, make a habit of following up on them regularly in your one-on-ones. A simple check-in like "Last time you said you wanted to get more confident in system design interviews. How's that going?" shows you're paying attention and that their growth matters.

You can also build light structures to keep them on track: note goals and progress in your shared one-on-one doc, set gentle reminders, or suggest small milestones. Even offering to review something they're working on or scheduling a future follow-up creates a sense of rhythm.

Both mentoring and coaching rely heavily on active listening: being fully present, asking thoughtful questions, and understanding what your team member truly needs. (See the section "Create a Safe Space" on page 97 for specific strategies on building active listening habits.)

Once you've created this space for reflection and exploration, one-on-ones become a great setting to uncover opportunities for delegation, which is a key way to support growth in practice. Pay attention to where someone is eager to stretch, or simply ask, "What kind of projects would you like to take more ownership of?" Then look for opportunities that align with their interests and goals. For instance: "I've been working on this initiative, but I think it could be a great opportunity for you. Would you be interested in leading it?"

One-on-ones also create a natural, lower-pressure environment for sharing feedback, both positive and improving points. You might say, "I really appreciated how you handled that conversation with the client, especially how you clarified expectations early on. It set a great tone." Or, when addressing something more difficult: "You mentioned wanting to get better at leading technical discussions. I noticed you stayed a bit quiet during yesterday's architecture review. How are you feeling about jumping in during those conversations?"

Making growth a priority in your one-on-ones is an investment in your team. Whether through mentoring, coaching, finding delegation opportunities, or providing feedback, your team members will feel supported and empowered to take on new challenges, which will speed up their growth.

Set Up Your One-on-Ones for Success

Before jumping into the strategies for setting up impactful one-on-ones, it's important to take a step back and clarify your role and make sure your team understands it too. In some teams, the tech lead also acts as the line manager. In others, there's a separate engineering manager or team lead who formally handles things like performance reviews and career progression. The split varies widely across organizations, and that's OK, as long as expectations are clear.

Even if you don't own formal responsibilities like compensation or promotions, you still play a huge role in your team members' growth. In fact, I'd argue that supporting that growth is part of your job. Your involvement might look different depending on the setup. For example, your conversations might not directly feed into performance reviews. Nevertheless, the mindset still applies: you are a key part of how your team develops. Chapter 7 goes deeper into how to contribute to reviews even if you're not the direct manager.

That's why the kinds of conversations described in this chapter are valuable in both contexts. Whether you're mentoring a junior developer, coaching someone through a tough technical challenge, or helping clarify next steps in someone's growth, you have influence. These one-on-ones are where much of that impact happens.

Now, let's look at how to set those one-on-ones up for success.

Making the most of your one-on-ones doesn't require overly complicated strategies. In fact, the biggest impact often comes from sticking to a few straightforward principles. But as simple as these may seem, many tech leads struggle to apply them consistently.

The foundation of successful one-on-ones starts with consistency. Whether weekly, biweekly, or monthly, they need a regular cadence. Sporadic check-ins won't cut it. Clear expectations are equally important; both you and your team members should understand what to expect from these conversations. Focused discussions help keep the meetings valuable by addressing the right topics. Trust and collaboration are built when the space feels safe for team members to share openly. Last, follow-through is critical: track action items and revisit previous discussions to ensure progress and accountability.

In the next sections, I'll break down each of these principles and share practical ways to implement them, so you can make your one-on-ones as impactful as possible.

STICK TO A REGULAR CADENCE

The single most important factor for successful one-on-ones is consistency. Yet, it's one of the areas where tech leads often fall short.

Regular, predictable one-on-ones create a foundation of trust. Whether you meet weekly, biweekly, or monthly, the cadence needs to be reliable.

I recommend meeting with each team member every two weeks for 30 minutes. For a team of up to eight people, this means scheduling four one-on-ones one week and four the next, which works out to just two hours per week, a manageable commitment.

Of course, the size of your team will influence what's realistic. My ideal team size for one tech lead is a maximum of eight people. This number makes it possible to keep up the biweekly schedule while staying deeply connected to the team.

For larger teams, maintaining this cadence can be more challenging. In those cases, it becomes even more important to be intentional about how you adapt your one-on-one schedule to make sure everyone still feels supported and seen. Here are some strategies that tech leads use in these situations:

Reduce frequency

Shift to a three-week rotation instead of two. Spread out the one-on-ones to avoid burnout while still maintaining regular touchpoints.

You can also ask people directly what cadence and duration works best for them. The problem might solve itself through conversation; some people you think need more contact may actually prefer less, and vice versa.

That said, reducing frequency can impact the strength of the relationship, so it should be a carefully considered trade-off. Just make sure to set the right expectations and align with your team before making any changes: explain the reasoning behind it and ensure everyone is on the same page.

Shorten the duration

Many tech leads default to 60-minute meetings, but half an hour is often enough when you stay focused. A shorter, consistent one-on-one is far more effective than an hour-long meeting that frequently gets postponed or canceled.

I generally advise against going below 30 minutes. Anything less tends to feel rushed and doesn't give enough space for meaningful conversation or deeper check-ins.

Adapt frequency by need

Prioritize newer team members or those who need more support. For example, schedule more frequent one-on-ones for new joiners and reduce them for senior members who are already well-established (as long as they agree with this).

Make temporary adjustments

When faced with tight deadlines, you might need to reduce or skip one-on-ones temporarily to focus on hands-on work. Use this option carefully and with caution! Always set clear expectations and communicate up front about the changes.

While it's normal for the frequency of your one-on-ones to evolve over time, consistency must remain the rule. Skipping a one-on-one should always be a last resort, and you should ensure it's a rare occurrence.

The consistent presence of these meetings on your calendar sends a powerful message: you value your team members, you're committed to showing up for them, and you care about the relationship you're building. That reliability is what establishes trust: the foundation for everything else you'll achieve together.

AGREE ON AGENDA AND EXPECTATIONS

For one-on-ones to work, everyone needs to know what to expect. It's up to you to set those clear expectations: what you expect from them and what they can expect from you. These meetings should feel intentional, predictable, and, most importantly, theirs.

Start by aligning on the purpose: these meetings are about supporting them, addressing their challenges, and helping them grow. When team members own the agenda, one-on-ones become more meaningful. Encourage them to take the lead, bringing their priorities, questions, or topics to discuss. If you have something to add, let them know ahead of time. When they drive the conversation, they're more likely to come prepared and open up.

To ensure shared ownership:

Let them drive the agenda

It's about their priorities, not yours. If they're unsure, guide them on what to bring to the table, but leave the floor open for their input.

Focus on their concerns

Discuss what's top of their mind. If they ask for help, ensure they feel safe doing so.

Be prepared as a backup

Have a few topics in mind in case they don't come with something specific that week.

Your first one-on-one is your chance to set the tone for future one-on-ones. Use it to explain the following:

What they can expect from you

Consistency, showing up prepared, and using this time to focus entirely on them.

What you expect from them

Taking ownership of the agenda, being honest about challenges, and using this space for meaningful discussions.

Topics of conversation

Check the section "Know What Topics to Cover" on page 94 for ideas.

Confidentiality

Unless there's a serious concern that needs to be escalated for safety or ethical reasons, what's shared in your one-on-ones stays between the two of you. If there's ever something that needs to be communicated more broadly, like to a manager or HR, you'll be transparent about it and let them know beforehand. The goal is to create a space where they can speak openly without fear of unintended consequences.

Here are some questions you can use:

- What do you expect from these one-on-ones?
- What do you expect from me as a tech lead?
- What would make these meetings valuable for you?
- How can I best support your work and growth?
- How do you like to prepare for these meetings?

This alignment builds trust and ensures one-on-ones are productive. A clear structure goes a long way in building trust and consistency:

Frequency

Set a regular cadence that fits the individual's needs. Some may benefit from more frequent check-ins, while others may prefer less frequent ones. Whatever frequency you agree on, make sure there is clear alignment.

Calendar invites

Make it official by sending recurring invites.

Creative naming

Reduce pressure with casual names like "Weekly Sync" or "How Can I Help?"

Description

Include a short note outlining the purpose of the one-on-one and any agenda points. To keep it simple, just link to your shared one-on-one notes document instead of updating the calendar invite each time.

Setting clear expectations from the beginning will reduce the pressure on you and the attendees that usually comes with this meeting.

KNOW WHAT TOPICS TO COVER

One-on-ones can cover a wide range of topics, and if you're doing it right, you'll rarely run out of meaningful things to talk about. From small talk to feedback to team dynamics, each conversation is a chance to connect, clarify, and collaborate.

For example, building relationships often starts with something as simple as small talk. Kicking off with "How was your weekend?" or checking in on something personal like "How's the home-buying process going?" can open the door to more meaningful conversations. You can also ask openers like "What have you been up to this week?" or follow up on something from your notes.

One strategy I've found effective for getting people to open up is starting with a genuine, honest answer to "How are you?" yourself. For example: "I'm really excited about a piece of work that's coming up" or "To be honest, I'm having a rough day today." Leading with openness gives permission for the other person to do the same. If all you're offering is "I'm doing OK," you're missing an opportunity to deepen that connection.

Growth is another essential area. Sometimes, team members naturally bring this up: "I've been working on improving my communication skills" or "I need help getting better at X." But when they don't, it's on you to make it part of the conversation. Ask questions like "What are your goals for the next few months?" or "What areas do you want to grow in, and how can I support that?" You can also go deeper with questions such as "What are your goals for this year?," "Do you have a plan to reach them?," or "What do you need from the team or the organization to support your progress?" This helps you understand where they are now, where they want to go, and how you can help them get there. You can also follow up on goals. Ask questions like "How is that communication course going?" or "Have you had the chance to try out that new skill in a project yet?" Showing you remember and care about their growth builds trust and keeps them accountable in a supportive way.

Bring visibility to their work. People often don't know how much their work is valued until someone tells them. Use one-on-ones to acknowledge contributions that might otherwise go unnoticed. You might say, "I think you handled that conversation on Tuesday with the client really well." It reinforces their impact and encourages continued initiative. You can also help them find other opportunities to make their work more visible, as this is an area many engineers struggle with. Ask questions that open the door: "That was a great training on Kafka Streams you gave to the team. How would you feel about presenting it to other teams as well?" or "Have you thought about writing a short internal blog post about the debugging process you figured out last week?" Small nudges like these can build confidence and help them develop their voice beyond the immediate team.

Feedback often fits naturally into one-on-ones. If there's something you've been meaning to share but haven't had the chance, this is the perfect time, though don't make the mistake of relying solely on these meetings for feedback. And it goes both ways. Use this time to ask for feedback from them: "How do you think I handled that situation?" or "What could I have done better in this morning's presentation?"

Team dynamics can also come into focus during these discussions. Ask open-ended questions like "What do you think we need more of in the team?" or "What's something you think we could do better?" These questions often reveal underlying issues or areas for improvement you might not have seen otherwise, and they open the door to something even deeper.

Use one-on-ones as a regular safety check. Beyond surfacing ideas and team improvements, these conversations are a powerful tool for gauging how safe and supported your team members feel. You can start by asking, "How are you feeling about the team these days?" or "Is anything making your day-to-day harder than it needs to be?" These kinds of questions help you monitor psychological safety and create a space where people know it's OK to raise concerns, big or small, before they become blockers.

Be mindful of changes in tone, energy, or participation. If someone who's usually engaged seems withdrawn or vague, gently check in. Questions like "You seem quieter than usual; how are things going?" can open the door for honest conversations that might otherwise be missed.

To build reflection into your one-on-ones, consider incorporating a few lightweight, recurring questions. These can help people pause and take stock of their week while giving you insights into their mindset and workload. Try asking:

- What's something you learned this week?
- What's one thing that went well?
- What's been challenging you recently?

Make space for wins. One-on-ones are a great opportunity to celebrate progress, even the small stuff. Asking questions like "What's something you're proud of from the last sprint?" or "Has anyone on the team helped you out recently in a way that stood out?" helps surface moments of success and appreciation. These quick reflections can be energizing and often trigger further conversations about motivation and what's working well in the team.

Pay attention to throwaway comments or subtle hesitations. "I've been dealing with some weird bugs" might just be a vent, or it might signal a larger problem worth digging into. One-on-ones are a great space to gently explore these threads.

Then there are the inevitable questions and worries that people bring. They might ask, "I heard there's going to be some restructuring. How will that affect us?" These moments are your chance to calm nerves, provide clarity, or at least show empathy when you don't have all the answers yet.

The most valuable one-on-ones tend to revisit a mix of themes: connection, growth, feedback, team health, and reflection. You don't need to cover everything every time, but having this mental map helps keep conversations balanced over time.

Ultimately, one-on-ones are there to support them in whatever way they need. They should feel like they own the meeting and you're there as a partner in their growth.

CREATE A SAFE SPACE

Creating a safe space in your one-on-ones is essential. Without it, meaningful conversations can't happen, and the real value of these meetings is lost.

One way to build this kind of space is to drop the formality. When I was working in an office, I'd sometimes invite people for a coffee outside during our one-on-one time, my treat. We'd take a short walk to a nearby café, chat casually on the way, and then settle on a bench in the park for a more thoughtful conversation. People loved it. That shift in environment and tone, stepping away from desks and screens, helped lower walls and made it easier to connect on a more human level. So if you have the chance to meet people face-to-face, take it! Even in a hybrid setup, I would plan a day at the office with my team and schedule my one-on-ones in person whenever possible. These moments of presence and connection often go a long way in building trust.

The next step is how you show up in these conversations. One simple strategy I use to set the tone is starting with a genuine answer to "How are you?" Instead of a surface-level response, I'll share something real: "I'm feeling excited about this new project" or "I've had a tough morning because..." This honesty invites them to open up, showing that it's OK to be authentic in this space. If you want people to open up, you have to open up first. Vulnerability is contagious.

I once mentioned in a one-on-one with a team member that I was feeling overwhelmed adjusting to a new process we were trialing. She immediately opened up about how she'd also been struggling but didn't want to seem negative. That moment turned our dynamic: it moved the conversation from polite check-ins to a more open and honest relationship. It reminded both of us that we weren't alone in navigating challenges.

By being honest about your thoughts and feelings, you make it easier for the other person to do the same.

Listening is the most powerful tool you have for building trust and creating a safe space. It's not just about hearing words; it's about understanding the person in front of you. Truly listening helps you develop empathy by seeing things from their perspective. When you do this, you connect on a deeper level and gain insight into what drives and challenges them.

Here's how to make your listening count:

Focus on the speaker

One of the simplest yet most powerful shifts you can make is reminding yourself: this is not about me. I even keep a sticky note on my monitor that says exactly that. To really focus, you need to eliminate distractions: turn off notifications, put your phone away, and be fully present. Especially while working remotely, I found this harder to manage. So now, before one-on-ones or important conversations, I turn on "Do Not Disturb," switch to full-screen mode, and keep a notebook nearby to offload distracting thoughts.

Embrace silence

If you want to listen more, you have to talk less. Early in my career, I felt the need to fill every silence or always have the last word. But I realized I was unintentionally crowding out others' contributions. Not every one-on-one will flow effortlessly, especially early on or with more introverted team members, and that's OK. Silence can feel awkward, but it's a crucial part of the conversation. It gives people space to gather their thoughts and build the courage to bring up something difficult.

To change my habits, I started defaulting to mute during remote meetings to give others room. One time, I forgot I was muted and started talking, only to realize the team had continued the discussion without me. They were doing just fine. That moment reminded me: my job isn't to steer every conversation but to create space for others to contribute.

Over time, I also learned to embrace silence instead of rushing to fill it. One trick that helped: count to 39 in your head. Most of the time, someone speaks up long before you finish counting, but even if they don't, that pause often leads to richer contributions than if you'd jumped in too quickly. You can say "Take your time" or simply sit with the quiet for a moment. Creating comfort with pauses helps make one-on-ones feel safer and more thoughtful.

Reflect and seek feedback

Reflecting back what you've heard is one of the most powerful ways to show you're truly listening. After someone shares something important, take a moment to summarize and check if you've understood correctly: "So what I'm hearing is that you feel your ideas aren't fully considered during technical discussions because the group often leans toward a different approach. Does that sound right?" This kind of reflection not only helps clarify their message; it also shows that you're making an effort to truly understand them.

But listening doesn't stop when the conversation ends. Becoming a better listener takes deliberate practice. After key conversations, especially one-on-ones or feedback sessions, I set aside 15 minutes to reflect. How well did I listen, on a scale from 1 to 10? Where did I drift off and why? Did I interrupt? I also make a point of asking for feedback directly, with questions like "What's one thing I could do to make you feel more heard?" or "How do you think I handle conversations when we disagree?" Over time, the answers to these questions have helped me refine not just how I listen, but how I lead.

Ask clarifying questions

Use open-ended questions to dig deeper and gain more context. For example, "Can you tell me more about why you feel that way?" or "What do you think would help in that situation?" These questions encourage them to share more while keeping the focus on their experience.

These strategies help you create space for open, honest dialogue, but they work only if you can stay present in emotionally charged moments too.

If a team member comes into a one-on-one clearly frustrated, venting about a difficult situation with the team or a recent meeting, your job isn't to fix it right away. In those moments, the most supportive thing you can do is simply stay with them. Don't interrupt, don't offer solutions, and don't jump to defend yourself if you feel the frustration is aimed your way. Just stay present and let them finish. I know how hard this can be; I've been there. It can feel like they'll never stop, like you need to respond. But I promise: they always stop. And if you've truly been listening, they'll know.

At the end of these moments, I often just take a breath and say, "I hear you. I'm sorry this has been so hard. What would you like me to do about it?" Sometimes, their response is "Nothing; I just needed to get it off my chest." That's the point: the sharing is for them, not for you. And asking this simple question helps you clarify their needs instead of guessing. Worst case? They tell you exactly what they want you to do, and you can decide how to act on it. But more often than not, the act of listening is the support they needed all along.

The openness people show in one-on-ones is a good measure of team safety. Hesitation, tension, or frustration might signal deeper issues. These insights can help you address problems early and build a safer, more collaborative team environment.

TRACK CONVERSATIONS

One-on-ones aren't isolated chats; they're part of a long, ongoing conversation. Growth doesn't happen overnight, and some challenges can't be solved in a single meeting. Consistency and follow-up are key to making these meetings impactful, and the only way to do it right is to track what's being discussed.

I used to keep an online note for each team member, dedicating a few minutes after every one-on-one to jot down key points: what we discussed, action items, and topics to revisit. During these meetings, I kept a notebook handy for capturing high-level ideas, allowing me to stay present. I tracked personal details like "getting a dog" or "new hobby," professional milestones or projects they were tackling, and follow-ups on development goals or challenges for future conversations.

Tip

Having a shared file for one-on-ones was a game-changer. It allowed team members to add their agenda in advance and track action items together, ensuring accountability on both sides.

Even with a shared document, I kept my personal notes for extra context. These were for things like remembering personal details, feedback I wanted to circle back on, or insights I'd want to bring up later. This combination of shared tracking and private notes helped me show up prepared and keep the continuity of our conversations.

Tracking also strengthens your follow-up game. When someone raises a concern, use your notes to keep track of actions you need to take and circle back. Give them an update, even if the outcome isn't ideal or you couldn't take action. Just closing the loop builds trust and reinforces the message: I heard you, and I followed through. It shows your team they can rely on you and that their concerns won't disappear into the void.

Tracking conversations shows your team that their time is valuable and that you're invested in their growth. It's also an opportunity to ask simple but powerful questions: "How can I make your life easier?" These small actions build trust and make sure your one-on-ones aren't just chats but actually help move things forward.

AVOID COMMON ONE-ON-ONE PITFALLS

One-on-ones are powerful but only if you treat them with the care they deserve. Unfortunately, some of the biggest mistakes aren't about what you don't do but about what you do without realizing the impact.

For example, showing up distracted, checking your phone, replying to Slack messages, or looking at another screen signals that this meeting isn't your priority. Over time, that erodes trust. The same goes for turning these sessions into complaint outlets. Sharing your challenges is fine, but when one-on-ones become a space for you to vent about other team members, stakeholders, or leadership, it puts your report in an uncomfortable position and undermines psychological safety.

Another common pitfall: dismissing concerns as "just complaining." What you hear as a rant might actually be a sign of deeper friction or misalignment. As a tech lead, your job is to listen with curiosity and look for patterns or problems behind the words, even if they're not expressed perfectly.

Also be mindful of dominating the conversation with your own updates or personal stories. A common trap is thinking you're being relatable by sharing too much about your own situation, especially when a team member is vulnerable. For example, they share something they're struggling with, and your instinct is to jump in with a similar story of your own. While well-intended, this can make the conversation about you, rather than helping them feel seen and supported.

If you find yourself doing most of the talking, pause and reflect: Is it because you're trying to fill silence? Are you uncomfortable with their discomfort? Are you unsure what to say, so you're defaulting to what's easy for you? These are all normal instincts, but they shift the focus away from the person who needs you to listen most.

One helpful self-check: after every one-on-one, ask yourself, "Did I walk away understanding more about them than they learned about me?" If not, it might be time to rebalance how you're using that time.

Finally, never underestimate the damage of inaction. If someone brings up an issue and you nod through it but do nothing, or forget it entirely, it sends a clear message: this space doesn't lead anywhere. When that happens enough times, people stop bringing things up.

Being aware of the common pitfalls helps you avoid the slow erosion of safety and connection.

Overcoming Common Challenges

One-on-ones can be incredibly impactful when done well, but they often come with challenges that can derail their value. As a tech lead, you may encounter recurring issues like struggling to find meaningful topics to discuss, sessions being canceled too often, or feeling like you have to solve everyone's problems. In this section, I'll explore these common roadblocks and provide actionable strategies to overcome them, so your one-on-ones can become productive, consistent, and meaningful conversations that benefit both you and your team.

NOTHING TO TALK ABOUT

Troubleshooting disengagement starts with identifying the root cause. Is it a team-wide issue, or is it just one individual?

If it's the whole team, the problem might be you. This often happens when there's a gap in how you've set up or are running your one-on-ones.

We can use Table 4-2 to troubleshoot.

Table 4-2. Common challenges that arise on one-on-ones and suggested solutions

Challenge	Suggestion
Team members seem disengaged in one-on-ones.	Start by having an explicit conversation in the one-on-one. Ask for their feedback on how valuable they find these meetings and what you might do to improve them.
	Also, ensure your one-on-ones are consistent. Sporadic meetings or skipped sessions signal they're not a priority, which can lead to disengagement.
Meetings feel unproductive and scattered.	Encourage them to come up with an agenda beforehand. Align on the purpose and value of these sessions to ensure both parties are prepared and engaged.
Team members rarely bring up meaningful topics.	Discuss potential topics in advance. Make it clear what's fair game for discussion, so they feel comfortable bringing up points that matter.
The team is holding back or not sharing openly.	Build trust by creating a safe space for conversations. If this persists, revisit strategies in the section "How to Build Strong Relationships" on page 62 to address trust issues.
Conversations feel stagnant and repetitive.	Follow up on action items and progress from previous meetings. Show continuity and commitment by revisiting prior discussions and tracking outcomes.

Addressing these challenges will help pinpoint where to start addressing the issue.

If it's an individual, dig deeper. Disengagement from one team member might hint at personal challenges or demotivation. I once had a team member bring no topics for weeks. When I finally asked about it, I found out she was frustrated about missing out on a project she really wanted to be involved with. We discussed other opportunities that excited her, and the problem vanished.

Even though it's their responsibility to bring topics, your team members might struggle, especially early on. Here's how you can help:

Track everything

    Keep notes on what they're doing well, areas for improvement, personal updates, and team dynamics. These observations can trigger meaningful conversations.

Prepare prompts

    If your team member struggles to bring up topics, have a few open-ended questions ready to trigger conversation. You can find a more complete list of ideas and question types in the section "Know What Topics to Cover" on page 94.

If a team member consistently shows up without topics, acknowledge it directly: "I've noticed you've been quiet in recent one-on-ones. Is there something on your mind, or is it just me?" This can lead to unexpected insights and opens the door to reminding them about the value of these meetings.

In my experience, when someone says they have "nothing to talk about," it's rarely the whole truth. It's usually a signal that something deeper needs attention. Addressing the root cause is how you turn disengagement into meaningful conversations.

GROUND RULES ARE NOT RESPECTED

At the start of your one-on-ones, you and your team members agree on the rules and commit to making these meetings valuable.

When things start to slip, it all comes down to reiterating those expectations and recalibrating together. Here are a few common issues that can arise, along with some ways to address them:

Lack of meaningful topics

    If team members rely on you to guide discussions, address it head-on: "For the past few weeks, we haven't had much to discuss in our one-on-ones. Why do you think that is?" Remind them of the purpose: "This space is for you to bring up concerns and questions or focus on your growth. How can I support you in using it better?"

You can also share your own experience to build trust and normalize their hesitation. For example: "I remember a period when I was having one-on-ones with my own manager and I didn't bring up much beyond surface-level updates. I didn't realize it was OK to talk about the things I was unsure about, like whether I was growing fast enough or whether I should speak up more in meetings. Once I brought it up, it completely changed how useful those conversations became for me."

Frequent cancellations

Frequent cancellations suggest these meetings aren't being prioritized. Approach the issue with curiosity rather than frustration: "I've noticed you've canceled a few of our one-on-ones recently. Why is that?" or "How can we adjust to make these meetings more valuable and ensure they fit into your schedule?" Reiterating the importance of consistency and the value these sessions provide can help reset their commitment.

Sometimes cancellations are due to pressures outside their control, caused by things like organizational changes, shifting priorities, or overloaded calendars. If it becomes a pattern, consider digging deeper: Is there a bigger issue at play? Are they being pulled in too many directions by stakeholders or leaders? If so, see how you can help, whether that's advocating on their behalf, adjusting expectations, or surfacing the problem more broadly. Repeated cancellations may be a symptom of something larger, and addressing it directly shows leadership and care.

Hesitation to voice concerns

If team members aren't using their one-on-ones to voice concerns or challenges, it might signal a lack of trust or hesitation. Create a safe space by reminding them: "This is your time to discuss anything on your mind, whether it's something you're excited about or something that's worrying you. If there's anything holding you back from bringing these up, let's figure out how we can address it together." Reiterate that confidentiality is part of how one-on-ones work. You can say something like "Unless something poses a serious risk or requires escalation, what we discuss here stays between us." Making this expectation explicit can go a long way in helping team members feel safe enough to open up.

The system works only if both of you stick to the process. These rules aren't just about their responsibilities; they apply to you, too. Let them know they can hold you accountable as well: "If I'm late, cancel too often, or don't follow through on what I promised, I want you to call me out on it. The system works only if we both stick to it." Even better, be proactive. If you slip, missing a meeting or coming unprepared, own it before they have to. Modeling accountability builds trust faster than asking for it.

By holding each other accountable to the ground rules you've agreed on, you ensure that these one-on-ones stay valuable, intentional, and productive over the long term.

FEELING PRESSURE TO SOLVE PROBLEMS

As a tech lead, it's easy to feel like you need to solve every problem that comes your way. The team looks to you for guidance, but you are not responsible for solving everything. Your role is to point people in the right direction, not to fix everything yourself.

An effective tech lead listens, empathizes, and supports but does not try to fix every issue. You are not supposed to! Trying to carry the weight of everyone's challenges will not only burn you out but will also prevent your team from growing.

Instead of jumping in with solutions, think of yourself as a guide, and use the following approaches:

Coach

Use powerful questions to help them find their own solutions. One way to start is by simply asking, "What do you need from me right now?" They may just want a sounding board, not a solution. Asking this up front helps you calibrate your support and shows that you're there to meet them where they are.

If they say they'd like your help figuring it out, then follow up with prompts like "What do you think might work here?" or "How would you approach this if you had no constraints?" These kinds of questions encourage ownership and growth while still making them feel supported.

Delegate

Recognize when someone else is better positioned to help. If it's a technical issue, loop in a subject matter expert. If they're looking for career guidance, suggest connecting with a mentor. Delegating wisely extends your impact without stretching yourself too thin.

Refer

If the issue is personal or sensitive and falls outside your expertise, gently point them to the right resources. Be familiar with what's available, like employee assistance programs, mental health services, or HR contacts, so you can guide them effectively when the situation calls for professional support.

It's helpful to show empathy, as long as you also hold clear boundaries. Not every problem brought to you will, or should, be yours to solve. If a team member expects you to fix something for them, you can respond with "I understand this is challenging, but let's explore how you can tackle it. How can I support you in finding a solution?" This approach shifts the responsibility back to them, empowering them to grow and take ownership of their challenges.

When you stop trying to solve all problems, you create space for your team to step up, learn, and take responsibility. It also protects your energy and ensures you stay focused on your core responsibilities.

Remember, being a tech lead isn't about being a superhero. It's about being a guide, a listener, and a supporter, helping your team find their own paths while keeping healthy boundaries.

Key Takeaway

One-on-ones are one of your most powerful tools as a tech lead, for managing day-to-day work and for shaping the culture and strength of your team over time. When done well, these conversations become the foundation for trust, growth, and alignment. They help you surface issues early, give people space to be heard, and support their development in ways that are thoughtful and personal.

What you bring to the table matters: your presence, your consistency, your ability to truly listen, and your willingness to follow through. You don't need to have all the answers. What matters more is creating a space where people feel safe sharing both their wins and their worries. When team members know they can count on that space, real connection and progress follow.

If there's one mindset shift to take away, it's this: strong relationships drive strong teams. One-on-ones give you a chance to invest in those relationships consistently and meaningfully.

So protect that time. Show up fully. And make it count.

5

Unlocking the Power of Feedback

A Note on Terminology

In this chapter, I'll refer to two types of feedback—*positive* and *constructive*:

Positive (or reinforcing) feedback
> This feedback highlights what someone is doing well, reinforces good habits, and builds confidence.

Constructive feedback
> This feedback focuses on surfacing blind spots, helping someone overcome challenges, or encouraging new behaviors.

Technically, both are constructive: they contribute to someone's growth. But in this chapter, when I say constructive feedback, I'll mostly be referring to improvement-focused feedback: the kind that helps someone see what to do differently or better.

As a tech lead, one of your primary responsibilities is helping your team members grow. There's a whole toolkit available to you with items like mentoring, sharing knowledge, and delegation, but none are as impactful as feedback. Yet, despite its power, feedback is often underutilized or mishandled.

I see it all the time: tech leads who shy away from giving feedback because they're afraid of damaging relationships or pushing cultural boundaries, or they're unsure how to approach it. They worry that giving constructive feedback will make them seem harsh or unkind or jeopardize trust they've worked hard to build. So they stay quiet, hoping the issue will resolve itself; often, though, it just tends to fester.

On the other hand, there are those who deliver feedback carelessly, leaving team members discouraged or questioning their abilities. The issue isn't that feedback doesn't work; it's that feedback, when done poorly, can backfire. It can harm relationships, damage careers, and demotivate someone entirely. But when done right, feedback can be transformative.

Even the smallest comments can have a profound impact. A simple "Great presentation today!" can boost someone's confidence, reassuring them they're on the right track. Constructive feedback, delivered thoughtfully and with clear examples, can save someone years of heading in the wrong direction. It can help someone see what they might be missing and encourage real progress. I've seen it happen time and time again, and I've experienced it myself.

I'll never forget a moment that changed the way I thought about leadership. My tech lead sat me down for a thoughtful conversation about how I was showing up at work, especially in my interactions with other team members. He walked me through a few examples, pointed out some patterns, and gently helped me see the impact I was having.

The main message he shared, though not in these exact words, was clear: I was lacking emotional intelligence, and if I wanted to lead tech teams effectively, I needed to start developing it.

Even though he delivered it with care and context, it still hurt. But it was also a turning point. It made me look inward and start focusing on how my behavior impacted others, rather than just fixating on metrics, results, and code. That piece of feedback was a game-changer, not just for my development as a leader but for my entire career, to the point that now I'm a huge advocate for soft skills in tech, and I am helping others develop theirs.

Or the time I told a senior developer on my team that his habit of constantly interrupting others was hurting their confidence and making them less likely to contribute. He had no idea the impact his behavior was having, but once I pointed it out, he was grateful and started addressing it. Over time, his relationships with colleagues improved, people's perception of him shifted, and even the quality of his work benefited. As his relationships deepened and trust grew, the other team members felt more comfortable engaging with him, giving him honest feedback, and collaborating more openly. This not only improved his awareness of how others perceived him but also gave him more opportunities to learn and refine his approach, ultimately elevating the quality of his contributions.

Feedback isn't just about what you deliver yourself; it's also about encouraging others to use it. Imagine two team members in a continuous conflict, and one of them comes to you frustrated and saying "Every time I speak in meetings, he cuts me off." My first response would be to encourage them to give feedback directly to the other person, explaining how having the courage to address things openly can not only help resolve the current conflict but also strengthen their relationship and reduce future misunderstandings. I would also offer to coach them through preparing the feedback and delivering it in a way that feels authentic and comfortable for them. Often, the other person isn't even aware of the impact they're having, and hearing it can lead to a meaningful change.

By teaching your team the power of feedback and encouraging them to use it regularly, you amplify its effects and create a feedback culture. In this kind of culture, people push themselves, and each other, to grow. Positive feedback acknowledges contributions and boosts morale, while improvement-focused feedback highlights gaps and offers a path forward. Making use of both types of feedback consistently helps build stronger relationships, improve team performance, and keep the team motivated. And best of all, this magic happens even when you're not around.

Giving feedback isn't always easy, especially when it's improvement-focused and touches on areas the other person needs to work on. It takes thought, care, and practice. Even with careful preparation, it can feel risky because you never know how the other person will respond. No wonder so many tech leads avoid it, putting it off until it's absolutely necessary! But delaying improvement feedback just because it's uncomfortable means missing an opportunity to help someone grow. It also increases the risk of the situation escalating if the behavior isn't addressed early. I cover this in more detail in the section "Understanding the Five Principles of Good Feedback" on page 117.

This chapter is here to make giving and receiving feedback less intimidating and give you practical ways to use it effectively.

How to Get Useful Feedback from Your Team

Getting useful feedback from your team isn't easy. It takes effort because it involves people being open, risking vulnerability, and possibly stepping into conflict. These are uncomfortable dynamics for most people, and in tech environments, where the focus often leans heavily on logic and execution, they can feel especially unfamiliar. Add the authority piece that comes with being a tech lead, and it's 10 times harder.

When it comes to positive feedback, people often assume you don't need it. "Of course you're doing a good job," they think. "That's why you're leading the team." They don't realize how valuable a simple acknowledgment can be for your motivation.

Improvement feedback is even trickier. People are often hesitant to share it with their tech lead because they're not sure how you'll react, or they worry there could be negative consequences for them. Or sometimes, they just don't know how to give feedback in a helpful way.

If you want feedback that helps you grow, whether it's encouragement or constructive critique, you need to make it happen. Build trust with your team, show them how valuable their input is, and guide them on how to give feedback that's actionable.

In this section, I'll share simple, practical strategies you can apply immediately to get the kind of feedback that makes a difference.

HOW TO ASK FOR FEEDBACK

The way you ask for feedback has a huge impact on what you get back. If you want feedback that's helpful and actionable, you need to put effort into the process and prepare for it. The key is to be clear, specific, and intentional about what you're asking for.

Start by being straightforward about the areas where you need input. Asking vague questions like "Do you have any feedback for me?" often leads to vague answers. Instead, focus on something concrete: "I've been working on improving my feedback skills. What do you think I could do to improve further?" or "I'm struggling with delegating effectively. How do you think I'm doing?" When you're honest about what you're working on, it signals to your team that you're open to feedback, and removes the awkward guesswork about what they think you need to improve.

If you're unsure about what specific area to focus on, you can ask for help identifying blind spots: "What's one thing I should start doing as a tech lead?" or "What's one thing I could do to better support the team?" These open-ended but targeted questions give people room to offer insights without feeling pressured.

To make feedback even more specific, use the 1-to-10 scaling system. For example, ask, "On a scale of 1 to 10, how good do you think I am at giving feedback?" If they rate you a 10, ask why: "What am I doing that makes you think I'm a 10?" If they give a lower number, ask what's missing: "What do you think I need to work on to reach a 10?" The number itself doesn't matter; it's all about the conversation it triggers. People find it easier to share their thoughts when they can anchor them to a scale, and it helps you pinpoint actionable steps for growth.

Who you ask for feedback is just as important as how you ask. Don't limit yourself to one group; seek input from everyone around you. You want to make sure you are getting input from both people who challenge you and those who cheer you on. Your team members can offer valuable insights about your day-to-day interactions, stakeholders can provide a broader perspective on how your leadership is perceived, and mentors can give you strategic advice.

When to ask also matters. Don't surprise people with feedback requests. Give them time to think. For example, mention in your one-on-ones on a given week that you'd like to have a feedback session next week, and share the topics you'd like their input on. Sending an email in advance can also help: explain why their feedback matters to you, outline the areas you want to discuss, and remind them to prepare. Avoid pulling people into an impromptu meeting with a vague "I just wanted to get some feedback."

Another good moment to ask for feedback is during offboarding. When someone is leaving the team, whether for a new role, a different company, or even just an internal move, they're often more reflective and more willing to share open, honest observations. Use that window to ask what worked well, what frustrated them, and how you showed up as a leader. A thoughtful offboarding conversation can surface insights you may not hear otherwise and help you improve the environment for the rest of your team.

Finally, treat feedback as a priority by scheduling it properly. Block time on their calendar, and include a note explaining the purpose of the meeting and the specific areas you'd like to discuss. This shows that you value their input and sets the tone for a productive conversation. A simple touch like this can make all the difference in getting the feedback you need to grow.

Asking for feedback intentionally creates the conditions for honesty, thoughtfulness, and trust.

HOW TO RECEIVE FEEDBACK

The feedback, both positive and constructive, starts flowing in. Now what? How you receive feedback matters just as much as how you give it. In fact, it sets the tone for how others approach feedback in your team. If you want people to be open to feedback and grateful for it, you need to model that behavior yourself.

Start with gratitude. Whatever feedback you're hearing, whether you like it or not, remember that someone has put effort into sharing it with you. A simple "Thank you for taking the time to share this with me" goes a long way. Acknowledging feedback doesn't mean you have to agree with it; it simply shows respect for the effort and the intention behind it.

Next, validate the feedback. This means making sure you understand it fully. Ask clarifying questions like "Let me see if I got this right. Are you saying that...?" or "What do you mean by that?" If the feedback is vague or unclear, ask for specific examples: "Can you give me an example of when this happened?" Context is everything when it comes to feedback, and understanding the specifics will help you decide how to act on it.

One time, I saw a tech lead dive headfirst into a complex conflict because she assumed what someone's feedback meant and jumped straight to action instead of taking the time to dig deeper. A team member had approached her with feedback: "I think you should code more as a tech lead." She took it to heart, immediately apologized, and started explaining herself. She then spent time reworking her schedule to increase her coding hours. But despite the adjustment and all the effort she put into it, the same feedback came up again from the same person in their next session.

This time, instead of rushing to fix it, she dug deeper: "Why do you feel that way?" The response was revealing: "Because my previous tech lead used to code way more." She pressed further, asking, "How do you think me coding more would help you or the team?" The answer caught her off guard: "I don't know, but you asked for an area of improvement, and this is what I came up with." Finally, she asked, "How much more coding do you think I should be doing?" The team member replied, "At my last company, my tech lead knew every piece of the code and even coded overtime."

That's when it hit her. The problem wasn't that she wasn't coding enough; it was a misalignment of expectations about her role as a tech lead. By rushing to action, she had been trying to solve the wrong problem. So, she changed her approach. She took the opportunity to explain what the tech lead role entailed at this company, how it differed from what he'd experienced before, and why coding wasn't her primary responsibility. She also shared the many other tasks on her plate that didn't allow her to code as much as the rest of the team. The team member thanked her for taking the time to clarify, and the feedback never came up again.

This story highlights the importance of validating feedback before reacting to it.

Think about how you approach architectural decisions or feature requests. You'd never jump in and start building something based on a vague prompt; you'd probably ask questions, clarify goals, and make sure you understood the problem first. The same principle applies here: asking clarifying questions helps you understand the real concern and respond in a way that's thoughtful and aligned with the actual issue.

Let's explore some common scenarios and how to handle them:

When positive feedback isn't new

If someone tells you something you're already known for, something you're a natural at, or one of your commonly recognized strengths, like "You're great at facilitating meetings" or "You're great at breaking down complex problems for the team," don't just brush it off. Instead, dig deeper: "Why do you think it's important for me to keep doing this?" or "How does this action help you or the team?" Even familiar positive feedback can provide valuable insight into the impact of your strengths and reinforce behaviors worth continuing.

When you don't like what you hear

It's natural to feel defensive when you receive feedback that stings or contradicts your self-perception. For example, someone might say, "You come across as dismissive in meetings." But you might not see yourself that way at all; you might feel like you're always encouraging input and exploring multiple perspectives. Comments like this can catch you off guard, and your instinct might be to explain yourself or push back.

But resist that urge. Instead, stay curious. Ask clarifying questions like, "Can you tell me more about this?" or "Could you give me an example of this behavior?" Even if you don't agree with the feedback, understanding the perspective behind it can be incredibly valuable. There may be a gap between how you see yourself and how others experience you, and that's exactly where valuable learning can happen.

Another common example that can trigger defensiveness is when someone gives you feedback about something you said or joked about that made them uncomfortable.

Let's say you made an inappropriate joke that didn't sit well with someone on your team. They let you know it made them feel uncomfortable. It's tempting to respond with "That's not what I meant!" or "It was just a joke." But doing that invalidates their experience and shuts down the conversation.

Instead, acknowledge what they're telling you and own the impact. You might say, "I hear you. I'm really sorry it came across that way. It wasn't my intention, but I understand it made you uncomfortable. I'll be more mindful next time."

In moments like these, your intention doesn't matter as much as the impact. You don't need to defend yourself; just take the feedback seriously, thank them for raising it, and commit to doing better.

When you're unsure how to react

They just throw some raw, unstructured, or completely unexpected feedback at you, like "I think you handled that situation really badly." It might come out of nowhere, catch you off guard, or be hard to make sense of in the moment.

You can't always control how or when feedback is delivered, but you can control how you respond. In this case, it's OK to say, "I need some time to think about this. I'll come back to you." This shows that you take the feedback seriously and gives you the space to process it without rushing into a reaction.

During the feedback session, avoid interrupting the person sharing their thoughts. If you have questions, jot them down and ask them once they're done.

Remember, receiving feedback doesn't mean you have to agree with everything. What matters most is staying open and curious and making an effort to understand the other person's perspective. You don't have to act on every piece of feedback, but by listening thoughtfully and engaging in a meaningful conversation, you'll build trust and encourage others to continue sharing their insights with you.

WHAT TO DO WITH THE RECEIVED FEEDBACK

To make the most of it, all feedback should be tracked and documented. This isn't just helpful for performance reviews; it's also a powerful way to track your progress over time. It allows you to see how others perceive your growth and areas of improvement.

One of the simplest methods I recommend is creating an online document titled *My Feedback*. Use this to record every piece of feedback you receive, including the context, who gave it to you, and the date. Keep it easily accessible so you can revisit it regularly and reflect on your journey. Personally, I believe it's best if you take the notes yourself during feedback sessions, even in a remote environment (small comments that will help you remember what's important; you don't need the whole transcript). While some prefer using AI assistants to transcribe conversations, I find this can make the exchange feel less personal and even inhibit open dialogue. Plus, knowing you're being recorded might make some team members hesitate to speak freely. Always ask your team what they're comfortable with before introducing tools like this.

Once you've collected feedback, the next step is to analyze it. Start by inspecting the feedback to identify what's useful. Ask yourself, "What's new information?," "What feedback has been repeated?," "What can I directly apply?," and "How valuable is this to me and my goals?"

Then, refine it into something actionable. Feedback like "You need to improve your communication" is too broad to act on. Look for patterns or ask clarifying questions to get to the root behavior. Maybe it's "You tend to speak too quickly in meetings" or "You don't summarize decisions before moving on." The more specific you get, the easier it is to take meaningful action.

From there, make a plan. For areas you want to improve, define what success looks like and outline next steps. That could mean adjusting a habit, practicing a skill, or setting up new feedback loops. Don't hesitate to ask for support if you're unsure how to implement the changes, from a peer, mentor, manager, or even an AI assistant. For example, you can prompt "Help me create a weekly plan to improve this feedback."

This does not mean you have to act on every piece of feedback. The goal is to identify the insights that resonate most and create a plan of action. Just make sure to close the loop with the person who gave it, especially if they took the time to share something thoughtfully. Acknowledge their input and share why you're choosing not to act right now, whether it's due to prioritization, context, or a different perspective. Leaving feedback unaddressed without explanation can unintentionally signal that their input doesn't matter, which can discourage future feedback.

After applying the feedback, it's essential to follow up. Monitor your progress on the areas you're working to improve and document what's working or what still needs adjustment. Schedule a follow-up conversation with the person who gave you the feedback to check how you're doing. An AI assistant tool can help you draft a plan for how to follow up, prepare questions, or even simulate a response to difficult feedback so you can practice. Prompt: "I received feedback that I tend to interrupt people in meetings. Can you help me come up with three questions to ask in a follow-up one-on-one and a short message to open the conversation?" This creates accountability and shows your commitment to growth.

Whether you're giving or receiving feedback, keep in mind that, as a tech lead, your actions set the tone for your team. How you handle feedback demonstrates to others how they should approach their own growth. By investing in this skill, you not only improve yourself but also create a culture that values continuous learning and development.

How to Give Useful Feedback to Your Team

Great feedback is all about balance: a mix of improvement-focused feedback and positive feedback. It's not just about pointing out where someone needs to improve or praising them when they've done well; it's about doing both consistently and with purpose. And just as trust is vital when receiving feedback, it's equally important when giving it. Trust grows when team members know you're honest with them, celebrating their strengths while also being brave enough to highlight areas for growth.

Let's be clear: balanced feedback doesn't have to follow the "feedback sandwich" approach. Personally, I find it inefficient. Wrapping constructive feedback between two layers of praise can dilute the message and often comes across as insincere. In practice, it tends to confuse more than it helps.

If you focus only on improvement-focused feedback, your team may begin to resent you. They might start seeing you as someone who only delivers bad news.

On the other hand, if you give only positive feedback, it can erode trust as well. People know they aren't perfect, and when they're not given areas to improve, your praise begins to feel insincere, or they might question your ability to observe the work they're doing. Over time, they might think, "Sure, I know I'm good at running retrospectives, but what should I work on?" Funny enough, too much positive feedback can actually hurt your credibility as a leader.

There's no magic formula for how much positive versus improvement-focused feedback you should give; it's not about numbers or metrics. Instead, it's about developing awareness of your tendencies and trying to find a balance that works for you and your team. Do you tend to focus only on what needs to be improved? Or are you also regularly acknowledging what's going well?

For example, I used to focus almost exclusively on improvement-focused feedback. I was always on the lookout for areas of improvement and didn't hesitate to share them. Then, during a performance review, the structure required me to provide positive feedback as well. The positive impact it had on my team was undeniable. I realized how much people valued hearing what they were doing well. From that moment, I started giving positive feedback more often and with the same intentionality as corrective feedback: tracking it, providing clear examples, and delivering it at the right time.

If you're someone who avoids constructive feedback because you're worried about how it might be received, start small. Pick one person and practice having those conversations. On the flip side, if you find yourself primarily delivering constructive feedback, challenge yourself to throw in some encouragement or praise, even if it feels obvious. It might be as simple as saying, "Great job facilitating that retrospective. The way you handled the conflict was spot on."

Once you experience the value of stepping outside your comfort zone, whether it's giving constructive feedback or positive feedback, you'll see the difference it makes in your team. For me, it became impossible to go back.

Balanced feedback also means ensuring that everyone on your team receives it equally. Avoid falling into the trap of giving all your praise to certain team members while neglecting others. Favoritism, even if unintentional, can create a toxic dynamic and make people feel undervalued.

In this section, I'll explore the core principles of good feedback and then dive into both positive and constructive feedback, exploring their unique challenges and incredible value when done right.

UNDERSTANDING THE FIVE PRINCIPLES OF GOOD FEEDBACK

Most people think that useful feedback is all about telling someone how to change their behavior. In my experience, advice is just a small part of what makes feedback truly effective. Figure 5-1 shows the five key elements that make feedback useful: giving it at the right time, being as specific and clear as possible, staying honest, and making it a continuous practice. In this section, I'll break down each of these criteria, with real examples and practical tips on how to apply them effectively.

Good feedback is:

🕐	Timely
◎	Specific
⚙	Clear
✋	Honest
🔄	Continuous

Figure 5-1. Five principles of good feedback

1. Good feedback is timely

The value of feedback diminishes the longer you wait to give it. The more time that passes, the less relevant and impactful it becomes. Yet, many tech leads hold off, waiting for what they think is the "right moment," which often ends up being during performance reviews. But by then, feedback like "The way you handled that client conversation six months ago was great" is far too stale to make a meaningful impact.

Also, delaying constructive feedback can lead to long-term consequences. Instead of addressing issues promptly, frustrations build up, and when the feedback finally comes out, it's often wrapped in passive-aggressive comments like "Should we just put Chris in charge of all meetings, since he talks the most anyway?" or blunt, hurtful remarks like "Why do you keep making the same mistake?" By this point, the problem feels bigger, harder to address, and more damaging to the relationship.

Here's a common scenario: imagine you've been silently hoping a team member will improve their code quality. Instead of addressing it early on, you drop hints in retrospectives or make vague comments like "We need to improve our testing." Months go by, and the same issues persist. Now you're forced into a difficult conversation: "Your tasks consistently come back from QA with bugs, and this has been happening for months. We need to open a performance improvement process."

At this point, the team member is likely shocked. "Why didn't you tell me sooner?" they'll ask, and they'd be right to. Early, clear feedback could have given them the chance to adjust their habits, avoid escalation, and regain your trust.

I've seen team members turn things around completely once they understood the problem and felt supported in working on it. The earlier you speak up, the more manageable and less painful the issue becomes, for both of you.

Don't miss the chance to give feedback when it's fresh and relevant, whether it's positive or constructive. Aim to deliver it soon after the moment occurs, ideally the same day or within a few days. If strong emotions are involved, give the situation space to settle, but don't let it slip away entirely. Acting while the context is still top of mind ensures your words have greater clarity and impact.

2. Good feedback is specific

The more specific your feedback, the more useful it becomes. Specific feedback gives people a clear direction for improvement and provides tangible points to address. It focuses on facts, which reduces the chance of defensive reactions and shifts the conversation toward solutions rather than debates.

Let's take a common situation: addressing code quality with a team member. A vague comment like "Your code is full of bugs" might seem like feedback, but it's far from effective. Instead of leading to improvement, it's likely to trigger defensiveness. The other person might respond with "That's not true; most of my code passes QA," and they'd have a valid point if you're referring only to their latest task. Without specifics, they're left in the dark about what you actually mean.

Now imagine instead saying, "The task you finished yesterday had missing tests and bugs that were caught in QA." This shifts the focus to facts that are hard to challenge, making it clear which task you're talking about. It opens the door for the other person to explain what happened. Perhaps they were overwhelmed, or it was a type of task they hadn't handled before. Whatever the case, being specific transforms the conversation into a productive discussion about solutions rather than a defensive argument about their overall performance.

Specificity also makes feedback fairer. In performance reviews or promotion discussions, vague generalizations can introduce implicit bias or lead to unfair evaluations. By grounding your feedback in clear, recent, observable examples, you create a more objective and equitable foundation for assessing performance. It ensures that everyone is evaluated based on what they've actually done, not assumptions, impressions, or stereotypes.

A general rule: be specific. Never assume people know what you're referring to; it's better to repeat yourself than leave them guessing.

3. Good feedback is clear

Good feedback is not just about what you say; it's about how you say it. Even if your message is specific, it can still fall flat if it's buried under vague or hesitant phrasing. Clear feedback means expressing your message directly, without hedging, downplaying, or confusing the core point.

Let's consider an example. Instead of saying, "I think you might have said the wrong thing this morning, and I think it might have happened last week too...maybe it was a misunderstanding?," opt for something direct: "This morning in standup, you said we have to integrate three services, but we need to integrate only two. I was concerned because I had shared the correct information with everyone yesterday, and it may have caused some confusion." The second version leaves no room for ambiguity; it pinpoints the issue, explains the impact, and sets the stage for a constructive conversation.

Avoid vague qualifiers like "maybe," "I think," or "sort of." They dilute your message and make it easier to ignore.

4. Good feedback is honest

Insincere compliments can do more harm than good. If you have nothing constructive or genuine to say, don't force yourself to come up with something. People can tell when feedback isn't sincere, and it makes people trust you less.

When giving feedback, focus on being kind, not just nice. Nice might mean telling a developer "Great job on the feature!" when you know the code wasn't up to the team's usual standards. Being kind means acknowledging their effort but addressing the issue honestly: "I appreciate the effort you put into delivering this feature quickly. However, the implementation could use some improvement, especially in terms of readability and edge case handling. Let's sit down and go through it together."

Kindness encourages growth and respect, while superficial praise only delays improvement and undermines credibility.

5. Good feedback is continuous

Feedback doesn't end when you give it. Tracking it is how you make it stick and see its impact.

Create individual notes for each member, jotting down the feedback you've given, the date, and any examples of progress or patterns that keep showing up.

Follow-up is where the magic happens. If the behavior improves, acknowledge it. Positive reinforcement goes a long way. If it doesn't, don't just let it slide.

Bring it up again with clear examples, and work together on the next steps. The intention is making sure it actually leads to real, lasting change.

Mastering these five principles, timely, specific, clear, honest, and continuous, is what turns feedback from a vague suggestion into a meaningful catalyst for growth. Whether you're praising a team member or helping them course-correct, these principles ensure your message lands with clarity and purpose.

HOW TO GIVE POSITIVE FEEDBACK

Most tech leads don't give enough positive feedback because they assume it's obvious. Saying things like "That was an awesome presentation, very clear and to the point," "I really liked how you handled that conversation," or even a simple, "I really appreciate you jumping in to help Ana with that bug yesterday" just doesn't come naturally. Tech leads often think, "They already know they're good at that" or "Someone else must have told them," completely underestimating the value of these words. But what if that "someone else" is thinking the same thing? Unlike improvement-focused feedback, which highlights areas for growth, positive feedback reinforces what's working well and motivates people to keep doing it.

A lot of people associate feedback only with criticism or corrective feedback, assuming positive feedback isn't necessary. Meanwhile, the techies I work with often tell me how much they crave their tech lead's acknowledgment, but they rarely get it. Positive feedback seems to show up only when something huge happens like finishing a project early, during a retro, when going above and beyond or staying late to fix an incident. And even then, it's often just a quick "Great job."

People need encouragement to stay motivated. They need proof they're heading in the right direction. Positive feedback is your chance to provide that, and it's just as important as constructive feedback. Genuine appreciation strengthens relationships: people feel more connected to those who recognize their efforts.

And positivity is contagious: when someone feels seen and valued, it lifts their mood and often spreads to others. A team that regularly celebrates wins, big or small, tends to be more engaged, collaborative, and resilient. It can be as simple as adding a kudos column to your retrospective board where team members can recognize each other's contributions. The emotional impact of positive feedback directly affects how people show up at work.

The key is to treat it with the same level of effort and intentionality. Apply the same five principles I introduced earlier in the chapter to positive feedback, and you'll see the difference it makes:

Timely

Positive feedback is most valuable when it's given as soon as possible. Don't wait six months to mention something great they did during a performance review. A simple "Awesome presentation today" or "I really appreciated how you handled that situation yesterday" during a one-on-one can have a massive impact.

It doesn't even have to be a formal conversation. Imagine someone nails a presentation. Send them a Slack message: "Loved how you presented the strategy using those visual diagrams. Great job showcasing the impact of our work!" A short, timely comment like this can make a big difference.

Specific

Vague praise like "You communicate well" is nice, but it doesn't help someone understand their strengths. Instead, pinpoint exactly what you appreciated: "The way you handled that tough conversation with the client the other day was impressive. Staying calm and finding a solution they agreed to really stood out." Specific feedback not only helps them see their strengths but also shows that you're paying attention.

Clear

Clarity adds weight to your feedback. Instead of saying, "I feel like you're ready to lead this initiative," connect it to specific examples: "The way you handled this task, dealing with dependencies, aligning the team on strategy, and delivering on time, shows you're the perfect fit to take over this initiative." Clear feedback leaves no room for doubt and boosts their confidence in taking on new challenges.

Honest

More positive feedback doesn't mean you should force it or say things you don't believe. If you don't genuinely have something good to say, don't say anything at all. People can see right through insincere compliments, and it will damage your credibility.

That said, if you put effort into noticing what people are doing, I am sure you'll find plenty of opportunities to praise them authentically. Everyone has things to be praised about also.

Praising effort, not just results, is also a powerful strategy, especially for team members who may not be the highest performers yet but are putting in real work to grow. Recognizing behaviors like perseverance, initiative, or thoughtful collaboration can reinforce those habits and keep people motivated. For example: "I really appreciate the time you've been investing in understanding the new system. It's clear you're putting in the effort to improve."

Continuous

Giving positive feedback isn't a one-and-done thing. Just like constructive feedback, you need to track it and reinforce it over time. If someone consistently demonstrates a skill you've praised, let them know you're noticing their growth: "You've been so consistent with your communication skills lately. It's making a big difference in how the team collaborates."

Even when applying these principles, many people struggle to come up with the right words for positive feedback. Here are a few examples to help you get started:

- I really appreciate how you structured that proposal; it was so clear and actionable.

- Thanks for stepping in during the incident yesterday. You kept things calm and organized.

- Pairing with the new joiners over the past two weeks has really helped them get up to speed quickly.

- Your documentation for the new process saved me so much time; thank you for putting that together.

- You've been doing a fantastic job leading standups; everyone's much more aligned lately.

- You consistently notice edge cases that others miss; it's a huge asset to our team's quality.

Keep in mind that how you deliver positive feedback matters too. While public praise can be a great motivator, in some cases it may unintentionally create discomfort or perceptions of favoritism. Some people simply prefer to receive recognition in private. Take a moment to consider the individual and the context: what's motivating for one person might feel awkward or alienating to another. When in doubt, tailor your message to what will land best with that person.

Positive feedback is one of the simplest, most effective tools you have as a tech lead to build motivation, confidence, and trust on your team. It doesn't require a big gesture, just your attention and intention. Make it a habit. The more you practice noticing and acknowledging what's working, the more your team will feel seen, supported, and empowered to keep improving.

HOW TO GIVE CONSTRUCTIVE FEEDBACK

There's a natural hesitation that comes with giving constructive feedback. Just like team members might avoid giving feedback to their tech lead for fear of repercussions, tech leads often fear how their team members will react. You might worry about hurting someone's feelings or damaging the relationship. And that's not something to dismiss. How you deliver the feedback matters just as much as what you say. If handled poorly, even well-intentioned feedback can cause defensiveness or resentment.

But avoiding the conversation isn't the solution, especially when the behavior is repeated. It only makes the problem bigger.

I once had a team member who consistently dismissed suggestions unless they came from her. Every time someone proposed a new idea, her first reaction was to poke holes in it, trying to find the flaw, the risk, the reason it wouldn't work. The rest of the team noticed, but our tech lead at the time said nothing. Over time, people stopped speaking up in discussions. They didn't want to risk being shut down, myself included.

Culture is not what we say. It's what we normalize. And in that silence, we normalized a culture where ideas weren't safe. By not saying anything, the tech lead unintentionally reinforced the behavior, and the damage spread.

A lot of tech leads use "I am too busy" as a reason to postpone tough feedback. But waiting doesn't save time; it creates more work. In this case, what could've been one honest conversation early on turned into several. The longer you delay, the more relationships are affected, and the harder it becomes to untangle the mess.

Some feedback cannot even wait until after. Sometimes, especially when incorrect information affects the direction of the conversation, you need to step in right away.

Let's say someone shares an outdated diagram during a client meeting, and the discussion continues based on incorrect assumptions. In those cases, it's better to gently interrupt: "Actually, it looks like this diagram is outdated. We've made a few changes since it was created. Let me pull up the latest version" or "We might need to revisit this once we've updated it."

It's a tricky balance, because while you don't want to embarrass anyone in the moment, you also have a responsibility to protect the integrity of the discussion. In situations like these, it's OK to step in to correct the information, especially when the conversation depends on it. What matters just as much is what you do after: follow up with the person privately. Check in on how they felt about you jumping in, and discuss how you can handle similar situations together in the future.

Addressing issues quickly keeps them from snowballing into bigger problems. It also shows your team that you're paying attention and that you care enough to address issues directly.

I'll go over some strategies that can make delivering constructive feedback easier.

Start with the five principles of good feedback we covered earlier: timely, specific, clear, honest, and continuous. If you keep these in mind, you're already ahead of most.

Adapt your style to the individual. Understanding the person you're giving feedback to is just as important as the content of the feedback itself. Each team member is different, and every new joiner brings a unique dynamic. Some may prefer directness; others may need more context or time to process. Pay attention to their communication style, personality, and experience level, and adjust your approach accordingly. You can even ask them how they prefer to receive feedback or which style they prefer.

Another strategy I've found invaluable comes from Nonviolent Communication, discussed in the section "Communicate Effectively" on page 63: focus on your own feelings. It's surprisingly effective, and underused. Usually, as adults, we are not encouraged to talk about how we feel, especially at work. I come from a culture where discussing emotions isn't really the norm. But once I saw how it shifted the tone of difficult conversations, both for me and the other person, I couldn't stop using it.

Saying something like "I felt dismissed when you jumped in over me in this morning's meeting while I was explaining how the system works" makes it hard to argue. If they respond with "I didn't mean to" or "It was just a misunderstanding," bring the focus back gently: "I understand it wasn't intentional, but that's how it landed for me," so they're aware of the actual impact of their actions.

Using a format to structure your message can have a big impact. It helps you stay focused, reduces emotional ambiguity, and makes your feedback easier to hear. AI assistants can support you here by offering different mental models or frameworks, like COIN or SBI. You can ask them to explain each model, compare their pros and cons, or help you choose the one that best fits your situation.

Let's say you're working through a tough moment or trying to clarify your thoughts before a conversation. You might prompt an AI assistant with "Here's what I want to say. Can you help me express it using the SBI framework?"

SBI is one of the most practical and widely used approaches for giving clear, actionable feedback.

The SBI model was originally developed by the Center for Creative Leadership, and it stands for Situation–Behavior–Impact. This method helps keep feedback grounded in facts, making it easier for the recipient to hear and reducing the likelihood of defensiveness.

Here is how the SBI model works:

Start with the situation

Describe the specific context where the behavior occurred. Be precise about when and where it happened and who was involved. This clarity ensures the recipient understands exactly what you're referring to.

Move to the behavior

Explain what the person did, focusing only on observable actions. What exactly happened? Avoid making assumptions about intentions or motivations; stick to what you saw or heard.

Finish with the impact

Share how the behavior affected you, the team, or the work. How did this impact you personally (what you thought and/or felt)? How did it affect the team, project, or organization (e.g., confusion, delays, morale)? Be clear and specific about why this matters.

Let's take as an example one of the hardest scenarios for tech leads: addressing a senior developer's negative behavior. Imagine a senior team member constantly interrupts others in meetings or dismisses their ideas. Over time, the team disengages, feeling undervalued and unmotivated. This situation is challenging because the senior developer might not realize the impact of their actions. You might also hesitate to address it, either out of fear of damaging the relationship or because their expertise makes it feel like you're questioning their authority.

Using the SBI model, you can structure your feedback like this: "During yesterday's planning meeting, I observed that you interrupted several team members when they were sharing their thoughts. This made me uncomfortable because it affected the flow of the discussion. I also noticed that after each interruption, some team members seemed less engaged and hesitant to contribute further."

This kind of feedback is clear, fact-based, and leaves space for the person receiving the feedback to reflect and respond constructively.

This message can be broken down using the SBI model, mapped onto the situation, behavior, and impact, as follows:

Situation

During yesterday's planning meeting...

Behavior

I observed that you interrupted several team members while they were sharing their thoughts.

Impact

This made me uncomfortable because it affected the flow of the discussion. I also noticed that after each interruption, some team members seemed less engaged and hesitant to contribute further.

Once you've delivered the feedback, give the person time to process it. Not everyone can immediately dive into next steps. Pay attention to their reaction and decide whether to continue the conversation now or follow up later.

When you're ready to move forward, shift the focus to understanding the reasoning behind the behavior. Was it intentional? A misunderstanding? Or is there an external factor affecting their performance? Use a coaching approach to help them identify solutions. Ask open-ended questions like "How do you think we can address this?" or "What steps do you think would help?" People are more likely to commit to a plan if the ideas come from them. Only step in with suggestions if they ask for your advice, and even then, let them choose the option that feels most achievable to them.

By delivering feedback promptly, focusing on one topic at a time, and structuring it with tools like the SBI model, you create a constructive environment for growth. It makes the feedback more digestible for the receiver and reduces the chances of defensiveness.

When it comes to constructive feedback, delivering it once isn't enough. Feedback is only as effective as the follow-up that ensures it's addressed and acted upon. After you've had the initial conversation, your work isn't over. You need to track how the situation develops, observe changes in behavior, and decide whether the feedback needs to be reinforced or adjusted.

The simplest way to ensure you're following up effectively is to track the feedback you give. Create a system, like an online note for each team member, where you record what feedback you delivered, when you delivered it, any specific examples discussed, a timestamp, the actions you agreed on, and what has happened since. This serves as a record you can revisit over time, helping you see if progress has been made or if patterns persist. Make sure to store this information somewhere private and secure.

For example, if you've addressed issues with late arrivals to standups, track their attendance over the next few weeks. Did the feedback lead to improved punctuality, or are there still recurring issues? Adding specific examples, such as "Arrived on time for every standup this sprint" or "Late three times this week," allows you to keep the discussion grounded in facts during any follow-up conversations.

Likewise, if the conversation was about code review engagement, you might note: "Reviewed 5 PRs this sprint and left constructive comments on each." Or if the topic was about responsiveness in tickets: "Replied to support queries within 24 hours consistently."

Regardless of the outcome, following up is mandatory to ensure your feedback is not being ignored and is being acted on. If the behavior improves, use positive reinforcing feedback to acknowledge the progress. For example, you could say, "I noticed you've been consistently on time for standups this sprint. I really appreciate the effort you've put into this. It makes a big difference for the team." Recognizing their efforts reinforces the desired behavior and motivates them to continue.

If the behavior hasn't improved, it's time to revisit the feedback using specific examples to re-address the issue. Here's where the tracking comes in handy. For instance, "I noticed you were still late three times this week. Let's discuss what didn't work and what worked since our last conversation." By tying the follow-up to observable patterns, you make it clear that the feedback is ongoing, not a one-time event.

Feedback isn't a one-and-done process. As you track and follow up, continue to adjust your approach as needed. If the same issue persists, consider exploring deeper reasons behind the behavior. Is there an external factor affecting their ability to change? Are they unclear on the expectations? Or do they need additional support to succeed? Asking questions and revisiting the conversation helps ensure they understand the importance of the feedback and have the tools to address it.

On the other hand, if the behavior improves and becomes consistent, you can begin to phase out the need for follow-ups, focusing instead on recognizing and reinforcing their strengths. The ultimate goal is to create a culture where feedback leads to growth, not just temporary fixes.

Constructive feedback only achieves its purpose when it's followed through. Without tracking and follow-up, you risk creating confusion or leaving unresolved issues. Following up demonstrates that you're invested in their growth and that the feedback wasn't just a passing comment but a shared commitment to improvement.

Overcoming Common Challenges

Even with a solid understanding of how to ask for feedback, deliver constructive feedback, and apply feedback principles effectively, challenges will still appear.

Tech leads often find themselves facing recurring issues like running out of feedback to give, struggling to get their team to provide honest and constructive input, or building a feedback culture within the team. These are normal pain points you will encounter in the process of feedback.

Each of these challenges requires its own strategies to address. In the following sections, I'll break them down one by one and share practical, actionable approaches to help you navigate these tricky situations.

"I HAVE NO FEEDBACK TO GIVE"

Every single day offers opportunities for feedback; you just have to pay attention. Once you become intentional about looking for it, you'll start noticing things. The truth is, everyone has strengths to be recognized and areas where they can grow. Your role as a tech lead is to observe, acknowledge, and act on these moments.

If you're struggling to find feedback, start by reflecting on the behaviors that trigger strong reactions in you. Are there moments that frustrate you or make you uncomfortable? For instance, maybe in planning sessions, the product manager always asks, "Who wants to work on this ticket?" and the same few people always jump in, while others rarely volunteer and end up with little to do. Over time, this kind of imbalance can lead to tension or disengagement within the team, and it creates more work for you as those left out frequently ask for new tasks mid-sprint. Even just opening the conversation with the manager by pointing this out can lead to a more inclusive task assignment process. Together, you might explore new approaches that ensure everyone gets equal opportunities to contribute.

At the same time, take note of the positive actions that might otherwise go unnoticed. Maybe a team member consistently takes on tasks no one else volunteers for, or someone always steps up to help a colleague who's stuck. Even when it's part of their role, these efforts deserve recognition. Highlighting these actions shows your team that you value and appreciate their contributions, no matter how small they may seem.

To ensure you always have meaningful feedback to give, start keeping a record of your observations. Jot down what you notice: when and where something happened, what was said or done, how it made you feel, and the impact it had on you, the team, or the clients. This helps you stay prepared for feedback sessions by building a history of examples. Over time, this makes it easier to follow up on positive behaviors you want to reinforce or areas that need improvement.

Once you start paying attention and documenting what you see, you'll quickly realize you're never out of feedback. The key is to look, listen, and be intentional. There's always something to acknowledge, whether it's to celebrate progress or address challenges.

"I CANNOT GET MY TEAM TO GIVE ME CONSTRUCTIVE FEEDBACK"

Getting constructive feedback from your team can be challenging. Fear of your reaction, worries about damaging relationships, or concerns about how their honesty might affect their growth often hold people back. Building trust is essential, as discussed in Chapter 3. Beyond that foundation, here are a variety of strategies you can use to encourage your team to provide you with useful feedback:

Display vulnerability

When it comes to feedback, vulnerability can be your most powerful tool. Openly admit areas where you know you can improve. For example, say, "This is a topic where I know I have room to grow. Do you have any ideas on what I should start or stop doing to get better at this?" When you show that you're open to improvement, it makes it easier for your team to be honest, as you've already done the hard part: acknowledging the area of improvement.

Use structured input

Share a list of role expectations and ask team members to highlight three things you're doing well and three areas for improvement. This clear framework makes it easier for them to provide actionable feedback.

Focus on specific situations

Reflecting on specific moments that didn't go as planned is a highly effective way to encourage constructive feedback. For example, you can ask, "What do you think about the way I handled that situation? What could I have done differently?" By acknowledging that the situation didn't go smoothly, you're signaling to your team that you're open to critique and actively seeking to improve.

This approach reduces the pressure on them to identify flaws, as the example is already on the table. If their response is overly positive despite a clear misstep, it might indicate a lack of trust or hesitation to deliver tough feedback, which is a sign that there's more work to do in building psychological safety within your team.

Ask for anonymous feedback

Provide an alternative channel for feedback, like an anonymous form. This can encourage those who feel hesitant to speak up directly to share their insights. To make these forms truly effective, include open-ended, thoughtful questions that invite reflection and specificity. For example: "What's one thing I should start doing to better support the team?"

That said, anonymous feedback has its limitations. You can't easily follow up to clarify or dive deeper. In small teams, it's often possible to guess who submitted the feedback, which can erode trust. And if overused, it might reinforce a culture of avoidance rather than openness. Use it as a supportive tool to get people sharing but not as a substitute for a proper feedback conversation.

Go first

There's often that awkward pause at the start of a feedback session: "Do you want to go first?" "No, you go." Instead of waiting it out, take the lead. Just say, "I can go first, if it's the same to you." Starting with your own feedback, especially something thoughtful and focused on growth, can make the other person feel more comfortable and open to sharing feedback with you in return.

Ask targeted questions

Sometimes the key to unlocking useful feedback is to make it as easy as possible for your team to respond. Frame your request in specific, manageable terms, such as "What is one thing I should improve at?" or "What is one thing I can do to better support you?" Another effective question is "What is one thing I should start doing as a tech lead?" These focused and direct questions simplify the process for your team, guiding their thoughts and making it less intimidating to share feedback.

Don't call it feedback

Sometimes, the word *feedback* itself can feel intimidating, especially in formal settings. Instead, approach the topic casually by asking about specific events. After a challenging conversation, you might say, "How do you think that conversation went?" Similarly, following a presentation, you could ask, "What did you think about my presentation?" By focusing on the situation rather than labeling it as feedback, you create a more relaxed and open environment, encouraging honest and constructive input without the extra pressure.

Getting useful feedback from your team won't happen overnight. It takes consistency, intention, and a willingness to show up vulnerably. These strategies can help you encourage your team to start giving you constructive feedback more openly and quickly, but there's no one-size-fits-all approach. Try them at different times, with different people. Each individual will respond differently, and it may take some experimentation to find what works best. What matters is that you keep trying. When your team sees that you genuinely care about growing, that you act on what you hear, and that their input has a real impact, they're far more likely to speak up.

BUILDING A FEEDBACK CULTURE ON YOUR TEAM

A team that embraces feedback bonds faster because it requires openness and vulnerability. Constructive feedback speeds up individual growth by providing clear improvement points. Positive feedback boosts morale, giving team members validation and reassurance that builds confidence and motivation.

Without a strong feedback culture, however, teams often fall into stagnation, misalignment, and quiet frustration. Problems go unspoken, growth stalls, and trust can quietly erode. Over time, the lack of honest conversations becomes a bigger blocker than any technical debt.

But even being aware of all its benefits, many developers aren't fans of feedback because it can feel uncomfortable and might even put them in a tricky situation with another team member. Also, proper feedback requires preparation and time that they just prefer putting into something else.

The strategies in this chapter will help your team feel more comfortable giving you feedback and make that feedback more useful and actionable. But if you want to build a true feedback culture, that's not enough. A strong team culture requires everyone to be involved: you don't want feedback flowing only between you and your team members but also among the team members themselves.

Here are some strategies that can help you to have this multiplying effect:

Start with you

One of the most powerful tools you have to shape your team's culture, especially around feedback, is leading by example. It's not just about telling people what to do but about showing them every single day. Consistency is key here. How you handle feedback, the priority you give it, and the effort you invest all set the standard for what you expect from others.

If you want your team to embrace feedback, start with how you approach it. Show them what asking for feedback looks like, how to receive it gracefully, and what to do with it afterward. Apply the same care and effort when you give feedback, whether it's positive or constructive.

Let your team see how you incorporate feedback. Share examples of when a piece of feedback led you to change something, whether it's how you run a meeting, communicate an idea, or manage a process, and what happened as a result.

Build psychological safety

For your team to consistently provide honest feedback, they must feel psychologically safe. This means creating an environment where they feel comfortable speaking up, sharing ideas, and expressing their thoughts without fear of negative consequences. They need to trust that their input won't harm their chances of promotion or damage relationships within the team.

Psychological safety is the foundation for open communication, and without it, feedback will always be limited or surface-level. I'll dive deeper into this critical topic in the section "How to Create Psychological Safety on Your Team" on page 188, but it's worth remembering that trust and safety take time and consistent effort to build.

Encourage feedback between team members

Feedback shouldn't flow only between you and your team members; it needs to happen among them too. For example, when two team members are in conflict and one comes to you frustrated, don't take on the problem as your own. Instead, encourage them to provide direct feedback to the other person. Help them prepare for the conversation by walking through what they want to say and how they might say it. Remind them to focus on the behavior and its impact, not on judging the person. You can even rehearse the conversation together and help them rephrase statements using the same principles we've covered earlier: be specific, clear, and honest, and avoid assumptions.

Educate your team on feedback

Your team may not naturally know how to give or receive feedback. Teach them what useful feedback looks like and share strategies to make it more effective. Demonstrate the importance of being specific, clear, and honest. Explain how feedback can help the team grow and improve, not just individually but collectively. You can do all this in an open session with your team about feedback.

Create opportunities to practice feedback

A great way to introduce this practice is through an exercise called Speedback. Speedback is a quick, structured way for teams to practice feedback. I first encountered it during my time at Thoughtworks, and I've used it in every team since.

Speedback is a simple and efficient method for exchanging feedback among team members. The process involves pairing team members for four minutes, during which each person has two minutes to give feedback to their partner. After the four minutes, the pairs rotate, and the process repeats until everyone has had the chance to exchange feedback with every other team member. This can be done either face-to-face or remotely using tools like Zoom breakout rooms.

To ensure the session runs smoothly, block time on everyone's calendar in advance, including 30 minutes of dedicated preparation time for them to put together the feedback, and share tips for giving useful feedback to set expectations. A straightforward structure works best: for example, team members can share two things the other person should continue doing and two areas for improvement. Or it can be as simple as one thing: "One thing I appreciate about working with you is..." and "One area where

I think you could grow is..." If possible, involve an external facilitator to handle the logistics, such as tracking time and managing transitions, so participants can focus entirely on the feedback exchange.

It's important to emphasize that Speedback is just the starting point for deeper conversations. The goal is to trigger follow-up discussions where participants can explore the feedback further, share examples, and define actionable steps. These sessions are particularly useful after completing a significant milestone, when joining a new team, or during periods of complicated team dynamics, but I advise you to run them monthly just to get into the habit.

Make feedback a constant process

As useful as Speedback sessions are to get your team started with feedback, they are not a replacement for ongoing feedback. Running one session a month doesn't mean you've established a feedback culture in your team. To embed feedback into your team's culture, it needs to happen consistently and naturally as part of your team's ways of working. Start by planning regular Speedback sessions to get everyone comfortable with giving and receiving feedback. But don't stop there. Follow up with broader conversations and encourage the team to keep those discussions alive.

Involve your team in sustaining this process. Ask for volunteers to help move the initiative forward, whether it's planning the sessions, tracking progress, or scheduling time on everyone's calendar. When the effort doesn't just come from you but also from other team members, people are often more receptive. It also signals that feedback is a shared responsibility, not just something pushed by leadership.

If you notice that feedback isn't happening as often as it should, address it in your team retrospectives. Reflecting on how feedback is exchanged is just as relevant and valuable a topic for a retrospective as discussing how a feature was delivered, since strong feedback practices directly impact collaboration, alignment, and, ultimately, delivery outcomes.

Track progress

Like any process, tracking and measuring the success of your feedback efforts is essential. How else will you know you are improving? Start small with metrics like the number of feedback sessions held, how many people are actively involved, and how ownership is being shared. Beyond the numbers, focus on the impact feedback is having on your team.

Ask yourself: "Are people acting on the feedback they receive, or is it being ignored?" or "Are your efforts making feedback a visible and valued part of your team?"

Look for signs that things are improving. For instance, team members might feel less stressed and more prepared during performance reviews because they've already received regular feedback throughout the year. This reduces the pressure of last-minute feedback prep and ensures there are no big surprises.

Different stages of your team's development might require different ways to measure success. The key is to focus on what works best for your team right now and adapt as you go.

Building a feedback culture requires a mindset shift and an ongoing commitment. The real payoff comes when feedback becomes second nature: when your team regularly gives and receives feedback, reflects on it, and acts on it without prompting. That's when feedback stops being a "thing you do" and becomes just part of how your team works and grows together.

Start small, be consistent, and keep showing up for the process. Over time, the ripple effects will show, in your team's collaboration, growth, and ability to tackle challenges.

Key Takeaway

Feedback is how teams grow...and it starts with you.

Creating a feedback culture means making feedback a natural, expected, and valued part of your team's daily rhythm. That includes asking for feedback regularly, receiving it with humility, giving it clearly and consistently, and creating space for others to do the same, not just with you but with each other.

It takes practice, courage, and care. But when done well, feedback strengthens relationships, accelerates growth, and builds a foundation of trust that will carry your team through every challenge.

You won't get it right every time, and that's OK. Just stay open, stay human, and keep the conversation going.

6

Delegating

Delegation is the process of assigning tasks or responsibilities to others while maintaining accountability for the outcome. It means letting go of work that others can take on so you can focus on your main priorities. But delegation is more than just offloading tasks; it's about working as a team, relying on others, and creating opportunities for growth.

For example, as a tech lead, you might delegate a feature implementation to a developer instead of coding it yourself. You provide context, set expectations, and offer guidance, but you allow them to own the execution. The key is balancing trust, oversight, and clear communication to ensure success.

The reason why delegation is a key skill to master as a tech lead is simple: you can't do everything yourself. Eventually, your workload becomes unsustainable. Your role shifts from being a great individual contributor to enabling your team to succeed. The most effective tech leads aren't the ones who take on every critical task themselves; they're the ones who build strong, autonomous teams that can execute effectively without micromanagement.

Despite its importance, delegation is one of the hardest skills for tech leads to develop. Many struggle with it, often hesitating because they fear losing control, worry about the outcome, or feel like it's faster to just do it themselves. Some even hold back because they fear they won't get the credit for work they delegate.

That's why this chapter is structured to cover everything you need to know to delegate effectively and with confidence:

- The benefits of delegation: how it helps you gain back time, grow your team (and yourself), and improve overall team performance.

- A step-by-step guide to delegation, from deciding what to delegate and who to delegate to, to choosing the right delegation strategy, setting clear expectations, and following up.
- The four most common fears tech leads face when delegating and how to overcome them.

By the end of this chapter, you'll have the tools to delegate with confidence, scale your impact as a leader, and build a stronger, more self-reliant team.

Benefits of Delegation

Delegation is one of the most underrated yet powerful skills a tech lead can develop.

When done right, delegation frees up your time to focus on high-impact work while also empowering your team members by giving them ownership and growth opportunities. Over time, this leads to a team that runs more efficiently, makes better decisions, and requires less hands-on oversight from you.

Each of these outcomes is worth unpacking on its own, so let's look more closely at the core benefits of delegation and how they show up in your day-to-day work as a tech lead.

GAINING TIME FOR YOU

One of the biggest struggles tech leads bring to me in coaching is time management. They feel constantly overwhelmed, juggling too many responsibilities: coding, one-on-ones, alignment meetings, firefighting, and performance reviews while still feeling like they're never doing enough. There's always something important that gets pushed aside. They end up chasing their day, moving from one thing to another without ever feeling in control.

Delegation is one of the most effective ways to offload some of these tasks and win back time.

When working with overwhelmed tech leads, the first thing I do is analyze their workload. We look at what's on their plate and identify tasks that don't require their direct involvement, tasks that could be handled by others in the team. The goal is to redistribute their responsibilities effectively. Once we identify what to delegate, the next step is making it happen (more on that in the section "Delegation Process: Step-by-Step" on page 143).

A classic example I see all the time: a technical, repetitive task that only the tech lead knows how to do. Maybe it's for an important client, and it involves a system or service that no one else fully understands. It's a 30-minute task every week. Easy, routine, predictable. You could do it with your eyes closed. You even do the task on vacation because, well, it's just easier to do it yourself. You've been meaning to automate it or, at the very least, document the steps to do it, but there's always something more urgent.

There are multiple problems with this approach. Besides wasting your time on something that could easily be delegated, you're creating a knowledge silo and increasing your team's bus factor risk. If you're suddenly unavailable, whether due to illness, an emergency, or simply being on leave, no one else can step in. Your team depends on you for this task, and no one else gets the opportunity to learn it.

Now imagine delegating it, not just passing the task on but giving someone else ownership of it, including finding a way to automate it. Yes, it might take some effort up front to train them, but the payoff in the long run is that you can remove it from your to-do list.

These "quick" 30-minute tasks are the real silent killers of your time. It's never just about the time spent doing them; it's about the impact of context switching. Studies show that it takes around 23 minutes to fully regain focus after being distracted. So those "easy" tasks might be stealing more of your time than you realize.

Letting go of some things on your plate doesn't mean you're doing less; it means you're focusing on what really needs your attention as a tech lead—the things only you can do.

By delegating effectively, you take back control of your time. You can finally plan your work, make space for strategic thinking, and stop feeling like you're constantly playing catch-up.

HELPING YOUR TEAM (AND YOURSELF) GROW

Besides freeing up your time, delegation is also one of the most powerful ways to help your team grow and build confidence.

One of the most common traps tech leads fall into (especially in smaller teams) is trying to stay on top of everything. They want to know the status of every feature, make sure everything is moving forward, and be involved in every decision. But this approach quickly becomes unsustainable.

A strategy that completely changed the way I worked at Thoughtworks was assigning feature leads. Instead of keeping everything on my plate, each feature had an owner. That meant another developer, not me, was responsible for making sure there was a clear plan, ensuring the team was aligned, tracking dependencies, creating stories, making progress visible, and identifying risks early. Their job wasn't to do all the work alone but to own the process, involve the team, and escalate issues when needed.

My involvement dropped significantly. I was still accountable for the outcome and needed to stay informed, but I no longer had to micromanage every step. Depending on the feature, I even encouraged them to handle stakeholder updates themselves. That way, they had full autonomy, and I stepped in only if something went wrong.

As they took on these new responsibilities, I could see them testing out leadership skills, managing complexity, communicating across roles, and making trade-offs, all in a safe, supported environment. These opportunities built not just their technical confidence but their leadership capabilities.

At the same time, what I didn't expect was that I often learned more from the process than they did. There's a saying that you don't truly know something until you explain it to someone else. The longer you do things a certain way, the less you question them. But when you hand a task over to someone new, suddenly, they start asking all the right questions. "Why do we do it like this?" "What if we changed that?" "Could this be improved?" Delegating opened the door to fresh ideas, challenges to the status quo, and improvements I never would have thought of. I was able to help others grow, and I was pushed to grow as well. Letting go means being open to the idea that your way might not always be the best way. And people will surprise you, if you give them the chance.

Delegation works best when it's rooted in trust. It's about giving people opportunities to grow and making sure the entire team shares the responsibility of delivering work. And the best way to build confidence in someone is to tell them: "I trust you to handle this." This single sentence can completely change how someone sees themselves.

Delegation gives others space to step up, experiment, and learn while also helping you grow as a leader who empowers rather than controls. If you want to build a strong, resilient team, learning to let go, strategically and supportively, is one of the most effective things you can do.

BOOSTING TEAM PERFORMANCE

Beyond individual growth, delegation also strengthens the team as a whole. As more team members step into ownership, the team becomes more collaborative, resilient, and efficient. People start supporting one another, spotting gaps, sharing knowledge, and solving problems together. Instead of relying on a single point of decision making, they learn to move forward confidently as a group. This shift positively impacts the overall performance of the team.

A great example of this is handling technical interviews. I worked with a tech lead who believed she needed to be involved in every single interview to ensure quality hires. She spent hours reviewing resumes, conducting coding assessments, and leading debriefs. While hiring is undeniably important, this level of involvement drained her time and slowed down the hiring process.

We worked on a plan: instead of being the bottleneck, she trained senior engineers to take over key parts of the hiring pipeline. She created a structured rubric for technical evaluations, standardized interview questions, and a clear debriefing process. Over time, she stepped in only when absolutely necessary. As a result, the team fully owned the process, and the overall performance of both the hiring process and the team improved dramatically:

- The hiring process moved faster because it no longer relied on just one person. The team could conduct multiple interviews simultaneously, keeping the pipeline flowing.

- Team members had a say in who joined, increasing alignment between new hires and the existing culture.

- Everyone learned something. Some engineers developed interviewing skills, while the tech lead gained insight into hiring bottlenecks and how to avoid them in the future.

- Onboarding time was reduced. With more time freed up, the tech lead could focus on refining the onboarding process, ensuring new hires could contribute faster and more effectively.

A high-performing team is one that functions well even when you're not around. A big part of being a tech lead is helping the team get there. That means taking what's in your head and sharing it with others.

One of the biggest blockers I see in tech teams is work that isn't ready to be worked on. People end up sitting around because the next tasks aren't refined enough. The natural reaction is that the tech lead and product manager take over pre-refinement, preparing stories in advance so the team can keep moving. It works but only for a while. Eventually, it becomes another bottleneck, another task weighing you down.

In my team, we made pre-refinement a shared responsibility. Instead of me always working with the product manager to prepare upcoming work, everyone took turns. It became part of the team's weekly work, ensuring that when sprint planning came around, stories were ready to go. No one was left waiting, and I wasn't scrambling to do it all.

But of course, getting there was a journey. It didn't happen overnight.

We started by laying the groundwork. We defined a clear structure for user stories, including consistent formatting, required context, and clearly written acceptance criteria. That gave everyone a shared understanding of what "ready" looked like.

We also leaned into pair programming, but not randomly. We put more thought into how we formed pairs for pre-refinement. One person would bring the technical context, while the other might be newer to the feature or to refinement itself or have less experience. That way, the task got done well, and someone also learned in the process.

Before anything was brought in front of stakeholders or the product manager, we would often run pre-refinement discussions as a team, diving into technical considerations and edge cases together. The goal of these conversations was getting stories ready and building shared ownership and technical judgment.

With time, people got the hang of it. The structure, pairing, and shared discussions gave everyone the tools they needed to contribute meaningfully. And as their confidence grew, my involvement became less and less.

Delegation means investing in your team's growth so you can step back, knowing things will still run smoothly. And someone not being ready yet is not a reason to avoid delegating; rather, it's your invitation to start preparing them.

By sharing knowledge and responsibility, you remove yourself as a bottleneck and enable your team to function independently. A good test of healthy delegation is this: can you take a real two-week vacation without needing to check in constantly? If the answer is yes, you've done it right.

Of course, you might argue that teaching others to take over tasks slows the team down in the short term. And that's true, at first. But in the long run, your team becomes more autonomous, less dependent on you, and better equipped to move faster and more efficiently without unnecessary blockers. Less time wasted waiting for you. More time for you to focus on what only you can do.

As a tech lead, you're always playing the long game. Delegation is how you win it.

Delegation Process: Step-by-Step

Now that I've covered why delegation is essential and the countless benefits it brings, like gaining back your time, growing your team, reducing bottlenecks, and ultimately improving team performance, let's move on to the how.

Delegating is about making sure the right people are handling the right tasks in a way that benefits both them and the team. When done right, it builds confidence, distributes workload effectively, and helps the team operate smoothly, even without you overseeing everything. But if done without care, it can create confusion, frustration, and more problems than it solves.

Some tech leads struggle more with certain aspects of delegation than others. Maybe you find it hard to decide what to delegate. Maybe you're unsure who should take on a task or you're worried the outcome won't meet your expectations. That's why breaking down the process step-by-step is so important.

In this section, I'll go deep into each step of effective delegation: identifying the right tasks, choosing the right person and the right delegation strategy, and setting clear expectations, all while maintaining accountability.

WHAT TO DELEGATE

Many tech leads struggle with feeling like they need to do everything themselves when, in reality, the opposite is true. The first step to effective delegation is figuring out what can be handed off, something that a lot of tech leads find surprisingly difficult. It often feels like everything they do is essential, but that's rarely the case.

A simple exercise to start is this: at the end of the week, write down all the tasks you handled. Then analyze: what are the things only you can do? This might include holding one-on-ones with stakeholders, defining the technical strategy for a company-wide cloud migration, refining technical stories with the product manager, or running performance reviews. These are tasks that often require your insight and full attention. Everything else is potentially delegatable.

And honestly, even those high-level tasks like defining the technical strategy for a company-wide cloud migration don't always need to be yours. Another developer on your team should be able to step into some of those conversations over time. And likewise, you can share the responsibility of refining technical stories with the whole team. If your absence causes a roadblock, then you've become a bottleneck. What happens when you're not around? Why can't someone else take over? The reality is, in many companies, tech leads are expected to be the go-to for everything, but that doesn't mean it has to stay that way.

Start small. Delegating simpler, repetitive tasks first will help you get comfortable with the process while showing your team you trust them. You can delegate both technical and nontechnical tasks, as you'll see in the coming examples.

Another great category of tasks to delegate are tasks you don't want to do. That might sound selfish at first, but often these are tasks you've simply outgrown. Just because you find something tedious doesn't mean someone else on the team won't see it as a great learning opportunity. What's routine for you might be new and exciting for them, and a great opportunity to develop new skills.

Onboarding new engineers is also a great example of a delegatable task. Many tech leads take full ownership of onboarding, walking every new hire through documentation, setting up their local environment, and explaining company processes. While it's important to ensure new team members are set up for success, this entire process doesn't have to be done by you.

Instead, create an onboarding buddy system where a senior or mid-level engineer takes responsibility for guiding new hires through their first few weeks. They can help with setup, answer questions, and provide technical context. You can still check in at key points, but by delegating the hands-on part, you ensure new hires get support while freeing yourself up for higher-level tasks. Plus, it's a great leadership development opportunity for the buddy. (For a deeper dive into how to set up an effective onboarding system, see the section "Onboarding Effectively" on page 175.)

Meetings are another easy win. I worked with a client whose company had recurring guild meetings on architecture, frontend solutions, and other topics. Instead of attending all of them, I found team members interested in those areas and had them go instead. Their job was to attend the meetings consistently, bring back insights, and flag anything that might impact our team. The outcome was highly effective: it saved me time, kept the team informed, and gave those team members a sense of ownership and visibility across the organization. It also created learning opportunities and helped them build internal networks. On top of that, by ensuring someone always showed up, we demonstrated to our client that we prioritized these conversations and cared about cross-team collaboration, without it always having to come from me.

If you're unsure where to start, the Eisenhower Matrix is a great tool to categorize your workload (check out the section "Time Management" on page 45 for details). It helps you separate what's truly urgent and important from what can be delegated, postponed, or dropped entirely.

And if you want even more ideas, ask your team. One-on-ones are a perfect space for this. Try: "Is there anything I'm doing that you'd be interested in taking over?" You'd be surprised at the answers you get if you just ask.

These are just some ideas to get you started. Once you begin thinking more intentionally about delegation, you'll find even more opportunities to free up your time while empowering your team.

WHO TO DELEGATE TO

Once you know what to delegate, the next step is figuring out who should take it on. The goal is to balance team growth, workload, and efficiency. Choose someone who not only has the capability but is also in a position to benefit from the opportunity. The right match leads to a smooth handoff. Otherwise, the task may end up right back on your plate.

The following sections cover some things to consider when choosing the person to delegate to.

Who is interested?

Motivation often matters more than prior experience. Someone eager to learn will put in the effort to get up to speed quickly. Instead of assuming who might be interested, ask.

For example, a tech lead I worked with always was the one to run sprint demos or showcases. I encouraged him to delegate the responsibility to different team members on a rotating basis by simply saying in a standup, "I usually run the sprint demo, but I'd love for someone else to take the lead this time. Who's interested?"

This not only distributes the workload but also helps developers improve their presentation skills, gain visibility, and build confidence in communicating with stakeholders. It also supports inclusion by creating visible opportunities that more junior or underrepresented team members can step into—opportunities they might not have asked for otherwise.

If more than one person expresses interest, that's a great sign. You can rotate the responsibility across people, pair them up, or take into account the criteria that follow, like skill fit or availability, to decide who takes the lead this time. Either way, interest is a strong signal that's worth nurturing.

One-on-ones are also a great space to identify interest. If you notice a team member looking to grow in a specific area, you can frame it like this: "I usually handle [task], but I think this could be a great learning opportunity for you. Want to pair on it this time?"

Who has the knowledge?

How urgent or high-impact is the task? If something needs to be done quickly and correctly with minimal support, delegate it to someone who already has the knowledge.

If no one else on the team can do it, it means you are a bottleneck. Start by pairing with someone. Have them document the process and try replicating the task when there's no pressure. Next time, let them take the wheel while you stay in the background for support.

For example, a tech lead was responsible for setting up CI/CD pipelines for every new microservice. It was slowing him down, and the team relied on him as the only person who understood the setup. Instead of being a bottleneck, he paired with another engineer, documented the steps, and let them handle the setup for the next service. Eventually, multiple people could set up pipelines, reducing dependency on him.

If working remotely and both agree, you could even record the pairing session or create a quick screencast walkthrough. It's a lightweight, low-lift way to capture the knowledge transfer and helps the person reviewing the task later to document the process more thoroughly. It becomes a reusable reference for the whole team.

Be careful not to always choose the most qualified person. This can create knowledge silos or overwhelm certain team members while others miss learning opportunities. For low-risk tasks with no urgent deadline, give them to someone who wants to build experience in that area. Yes, it might take longer the first time, but in the long run, it distributes knowledge across the team.

What's their current workload?

Delegation is meant to reduce your workload, not just shift it onto someone else. Before assigning a task, check in with them. Some team members might say yes to everything because they see it as a privilege, but they may already be overloaded. Ask directly: "How would this fit into your workload right now?" instead of assuming that it does.

A common pitfall I've seen is tech leads always delegating to the same person. One tech lead always relied on a senior engineer for documentation-heavy tasks like writing architecture decision records (ADRs). After a while, the engineer got frustrated and asked why no one else was responsible for it. The tech lead realized that she hadn't considered workload balance and started rotating this responsibility, giving others a chance to develop that skill.

Does their working style match the task?

The right person isn't always the most experienced; it depends on the nature of the task.

If you need someone to own communication and alignment between teams on a high-impact project, pick someone who enjoys collaboration and has strong soft skills.

If you're trying to debug a complex memory leak, delegate it to someone who thrives on technical deep dives and enjoys solving intricate problems.

If you're unsure how to spot these traits in your team, you're not alone. It takes time to develop this skill. A good first step is simply paying attention to how people behave in different situations.

For example, someone who's a strong communicator often volunteers to clarify points in meetings, follows up in writing, facilitates retrospectives, and checks in with team members to keep everyone aligned. They tend to be proactive, taking on extra responsibilities, stepping up to lead initiatives, and looking for ways to be involved in broader conversations.

In contrast, someone who thrives on deep technical work might light up when tackling complex problems, eagerly jump into debugging, or take the lead during incidents. They're often the ones suggesting technical improvements or exploring new tools and approaches.

Pay attention to what energizes each person, what they naturally take ownership of, and how they engage with others. These patterns can give you valuable insight into their strengths.

Of course, this doesn't mean the task always needs to match their current style. Delegation can be a great growth opportunity. Sometimes it's worth giving a strong IC a chance to lead a cross-team initiative or asking a great communicator to dive deeper into technical ownership. These mismatches can challenge people to build new muscles and grow in unexpected ways, especially when they have deep expertise in a certain area and can bring unique value to the task.

These are not a strict checklist; you don't need to tick every box for delegation to happen. Often, you're balancing a mix of factors. You might not have someone who does not have the knowledge to do the task yet, but they're eager to learn. Or someone may be the ideal person, but they're currently at capacity. Or no one might be volunteering at all.

When no one steps up, try the following strategies to get people engaged.

Ask directly, especially if you think someone is a strong fit: "I think you'd be great at this. How would you feel about taking it on?"

Connect it to their growth goals: "In your last review, we talked about building confidence through knowledge sharing. Want to run the next show-and-tell session?" Explain why you think it's a good opportunity and have an open conversation. If they hesitate, explore why. Maybe they're unsure how to start. Offer support by pairing with someone or doing a dry run.

Ask for help, with vulnerability: "I'm overloaded with this stakeholder presentation. Could you please prepare and run next week's product demo?" Being honest about your own challenges makes people more willing to step up.

Don't expect everyone to jump in. But one person might, and that's enough to start momentum. Often, delegation creates compounding effects. When others see peers take on new responsibilities and succeed, they're more likely to follow. Highlight these examples in one-on-ones to encourage others.

The tone of your ask matters. Avoid assigning or pressuring people into something they clearly don't want. Ownership can't be forced.

Still, if you've tried all this and the team continues to resist, that's likely pointing to a deeper issue. At that point, I'd bring it to a retrospective. Talk about the elephant in the room—the lack of engagement or willingness to take ownership—and work together to understand why. Whatever the reason, surfacing the issue is the first step to solving it.

Choosing the right person to delegate to takes intention, not perfection. There's no one-size-fits-all answer, just a mix of judgment, context, and experimentation. The more you practice, the better you'll get at reading the room, matching people with opportunities, and creating space for everyone to grow.

CHOOSE THE RIGHT DELEGATION STRATEGY

Delegation doesn't mean dumping your work onto someone else. You can't just hand off a task and expect it to be done perfectly without any guidance. Some people will need more support, some less. Your job is to provide the right level of direction based on the person's experience, skill level, and confidence with the task.

Once you've identified the right person, the actual handover begins. Start by explaining why you're choosing them for this task. Maybe it's because they've shown interest in the topic or because of their technical knowledge, their ability to think outside the box, or their strong collaboration skills. Whatever the reason, say it out loud. This helps set the right expectations and avoids misunderstandings like "Why am I getting more work?" or "Are they just throwing this on my plate because they don't want to do it?" If people understand why you're delegating something to them, they're more likely to approach it with the right mindset.

Of course, give them a chance to ask questions, raise concerns, or even push back if it's not something they feel ready for or interested in.

Match delegation style to the situation

Not everyone needs the same level of support when taking on a task. The way you delegate should depend on the person's knowledge, experience, and confidence in handling it. Think of delegation as a spectrum, from high direction and hands-on support to full autonomy.

This approach is inspired by the Situational Leadership Model (originally developed by Hersey and Blanchard), which emphasizes adapting your leadership style to the development level of the person you're supporting.

Here are four key approaches based on the person's situation:

Telling (directing)

> This is for someone who has little to no experience with the task and may also be hesitant or unwilling to take it on. In this case, clear instructions and close guidance are necessary. Define the task step-by-step, explain exactly what's expected, and check in frequently to make sure they're on the right track.
>
> Example: A team member is preparing for their first client-facing demo. Instead of just handing over the calendar invite, help them understand the audience and their expectations, review the content together, and do a dry run where they can practice with you and get feedback. Attend the demo to offer backup if needed, and debrief together afterward.

Selling (coaching)

This approach works when someone is eager to take on a task but lacks the skills or confidence to do it on their own. They might hesitate because they think they're not capable. Here, your job is to build their confidence while providing the necessary training and support.

Example: If a team member is interested in leading sprint planning but feels unsure, you could coach them through the process. Have them shadow you the first time, then let them lead the next session with your support, offering feedback and encouragement afterward.

Participating (supporting)

At this stage, the person has the skills and experience to perform the task but may benefit from occasional input, encouragement, or collaboration. The focus here is less on instruction and more on confidence-building, motivation, or being a sounding board. You take a more collaborative approach, supporting their decisions, removing blockers, and checking in as needed while allowing them to take the lead.

Example: A developer who has previously worked on tech design documents together with you now wants to write one independently. Instead of reviewing every line, you provide feedback only if needed and encourage them to refine their approach. Finally, before the document is officially finalized, you might want to review it with them, ensuring alignment and making any necessary adjustments together, as the accountability remains with you as the tech lead.

Delegating (hands-off)

When someone is fully capable and confident in handling a task, your job is to step back. Let them own it entirely while staying available if they need support.

Example: A senior engineer has been leading technical discovery sessions for new projects. She is experienced in gathering requirements, assessing technical feasibility, and facilitating discussions with stakeholders. At this point, she doesn't need your guidance, just the space to do her job. You check in occasionally to offer support or remove blockers if needed, but you fully trust her to handle it independently.

Adjust your delegation approach

People don't stay in one category forever. The goal is to help them move from needing a high amount of direction to full autonomy. If you're always in "telling mode" with a capable team, you'll be micromanaging. If you immediately take a hands-off approach with someone new to a task, they might feel abandoned and struggle.

A good rule of thumb is to start with more guidance and gradually pull back as the person builds confidence and skills. Effective delegation removes work from your plate while helping your team grow and ensuring tasks are completed successfully.

It's useful to mention that not everyone needs the same level of support for every task. One person might be highly autonomous when working on backend systems but need more direction when dealing with stakeholders. It's not about labeling people but about adapting your approach to each situation and skill set. Think of it as calibrating your support to match the task at hand.

A great way to refine your delegation approach is by asking for feedback. After the task is completed, check in with them: how did they feel about the process? Did they get enough support? What could have been done differently? Their input will help you adjust your style and improve future delegations.

Effective delegation is about thoughtful adaptation. The more you tune into each person's experience and confidence level, the better you can support their growth and ensure successful outcomes.

SET CLEAR EXPECTATIONS

"I'll just do it myself next time." If you've ever caught yourself thinking that after delegating a task, unclear expectations might be the real issue. Clear, up-front communication is what separates effective delegation from a frustrating cycle of rework and misalignment.

I remember this time when I asked a developer to prepare a showcase presentation. I thought it was obvious: prepare slides, do a demo, and get ready to present for five minutes like we usually do. I told her to ask me if she had any questions, have everything ready before the showcase, and leave some time for feedback. She didn't reach out during the preparation process, so I assumed things were going smoothly.

Two hours before the showcase, I checked in to ask about the status, as I hadn't heard anything from her. She told me she had everything ready to go. Then, she showed up at the showcase with only one slide and one picture, which didn't take her longer than one minute to present. I was shocked. I definitely didn't react well at the moment. She didn't understand what she had done wrong, and to be honest, she was right. Because I never clearly told her what exactly I expected, I just assumed she knew.

That's when I learned a valuable lesson about the power of setting clear expectations when delegating. It's not enough to assume the other person understands. I needed to clearly communicate what I expected the outcome to be, define our roles in the process, explain how I'd know the task was done, and agree on all of this up front. After that experience, I made it a point to be much more specific about expectations whenever I delegated a task.

That's also when I started leaning on a familiar framework: SMART goals. SMART goals aren't just for personal productivity; they're a powerful way to create alignment and accountability when delegating. They help ensure both sides are on the same page about what success looks like, what's expected, and when it should happen.

But over time, I realized something was missing. Delegation doesn't stop after the initial handoff; you also need a way to track how things are going. So I added one more T to the model: trackable. The result is a lightweight but powerful extension I use with my teams: SMARTT goals. Let's break it down:

Specific

> Be specific in outcome, flexible in execution. Be crystal clear about what the expected outcome is. If needed, provide some guidance on how you expect the task to be tackled, but keep in mind that the person taking it over should have the autonomy to decide the best approach. If the task requires collaboration, specify who else should be involved: team members, other teams, or even different departments. Also, share any relevant information you already have to give them a head start.

Measurable

Define how success will be measured. This could be quantitative (e.g., reducing build time by 20%) or based on deliverables (e.g., producing a report with recommendations for improving system performance). Having a measurable goal helps track progress and ensures alignment.

Achievable

Is the goal realistic given the timeline and available resources? This is something you should discuss and agree on with the person taking over the task. If it's not achievable as originally planned, adjust accordingly.

Relevant

Why does this task matter? Even if you think the purpose is obvious, take a moment to reiterate the "why." Understanding how their work contributes to the bigger picture will make them more engaged and invested in the task.

Time-bound

Agree on a deadline. A clear timeline prevents tasks from dragging on indefinitely and helps the person prioritize effectively. Without this, things tend to get pushed to the bottom of the list, especially in busy teams.

Trackable

How are you going to keep track of progress? Will it be on a Confluence board or a shared doc or via a regular check-in? How often will you review progress together to ensure things are on track? Having a clear system for tracking progress and checking in on progress significantly reduces the risk of unexpected surprises down the line.

Table 6-1 shows an example of how you can apply SMARTT goals to a classic delegation scenario: delegate the initiative of reducing tech debt in your codebase to a senior engineer on your team.

By setting clear expectations like this, you significantly reduce the risk of misunderstandings, misalignment, and disappointing outcomes. When both you and the person you delegate to are on the same page from the start, it increases the likelihood of success and makes delegation a much smoother process.

Table 6-1. Example scenario using SMARTT goals

	Suggestions	Example: Delegate the initiative of reducing tech debt in your codebase to a senior dev in your team.
Specific	What are the expected results? Who should be involved? Provide all relevant information you have about the task.	You are responsible for leading the initiative to address and reduce technical debt within our codebase. This includes: • Reviewing the current list of technical debt tasks • Working with the team to identify areas of high technical debt • Prioritizing them based on impact and urgency • Creating a detailed plan for addressing these issues • Ensuring the team is following the plan
Measurable	Quantities/ deliverables	The goal is to address 30% of technical debt tasks over the next three months.
Achievable	Can the objective really be accomplished?	This tech debt task will be addressed alongside regular feature development tasks, with dedicated time allocated in each sprint.
Relevant	What is the point of doing this task?	Reducing technical debt is critical for: • Improving the long-term maintainability of our codebase • Speeding up future development • Reducing the occurrence of bugs • Reducing onboarding time for new developers This initiative aligns with our broader goal of ensuring product stability.
Time-bound	Clear timelines	This initiative should be completed within three months.
Trackable	How will progress be monitored or reviewed?	• Progress will be tracked in a shared Confluence page, updated weekly with completed tasks and blockers. • We'll review the status together during our Monday check-in meetings to ensure we're on track, and adjust if needed. • The plan and updates will also be visible to the whole team for transparency.

FOLLOW UP

Once a task is completed, regardless of the outcome, following up is essential. It supports your team's growth and helps you refine your leadership approach.

Share your feedback on how the process went, looking at both the final result and how things went overall with the person you delegated to:

- What worked well?

- What challenges did they face?

- What could be done better next time?

Make sure to ask them for their feedback too. A simple but powerful question like "What could I have done differently to support you better in this process?" keeps delegation iterative and continuously improving, making it easier for both you and your team to refine how you work together over time.

Follow-ups also help validate your perspective on how things went, because sometimes, what you think happened and what the other person experienced can be completely different. Here's an example: a tech lead delegated the task of documenting the deployment process to another developer in the team. The goal was to get a high-level view of all the steps involved so the team could later discuss and identify blockers or areas for improvement. The tech lead also asked that the document be created in Confluence, since that was the team's standard documentation tool.

The final result was great. The tech lead was happy with the work and assumed the other developer was too. But when they had the follow-up conversation, he was surprised to hear "I would have preferred using a better tool for building the diagram instead of Confluence, something more interactive."

Turns out, while the other developer did the task as requested, she was a little frustrated with the process but didn't say anything at the time. The tech lead realized he had been unnecessarily rigid about the tool choice. He wasn't even aware that this had caused friction until the follow-up chat.

As you can see from the previous example, delegation doesn't end when the task is done. Following up is an opportunity for growth and refinement, both for the person you delegated to and for yourself. It ensures that delegation remains a learning process rather than just a one-time transfer of responsibility.

STAY ACCOUNTABLE

Delegation is the process of assigning tasks or responsibilities to others while maintaining accountability for the outcome. This means that, as a tech lead, even when you delegate a task to someone on your team, you are still the one stakeholders will turn to if things go wrong. You don't get to say, "Well, that was someone else's responsibility." Just as you get credit for your team's successes, you also take ownership of its failures.

This is where the difference between accountability and responsibility becomes important:

Responsibility means taking ownership of the work.

Accountability means taking ownership of the outcome.

A tech lead is accountable for the team's outcomes, while the whole team is responsible for executing the work. Think of it this way: when things go south, the tech lead is the first one in the line of fire. That's why effective delegation isn't just about handing off tasks; it's about ensuring that the work is done right, without micromanaging.

Being accountable also means you can't suddenly offload your entire plate and expect everything to run smoothly. You can't just toss tasks at people and walk away. You have to support them in delivering it well. It's a balance; you need to delegate enough to empower others without disappearing from the process.

If you start throwing things into people's laps, not only will your team likely grow frustrated, but the work itself might suffer. And since you're still accountable for the outcome, it could easily backfire on you. Like you've seen throughout this chapter, even if you're not doing the task yourself, successful delegation still takes effort and intentionality. Start small by delegating one task at a time, so you can build trust, learn what works, and grow your confidence along the way.

The key to maintaining accountability while delegating is to create a clear feedback loop. Regular check-ins, structured updates, and clear expectations (as I cover in the section "Set Clear Expectations" on page 151) help them do a great job.

Of course, even with all of these practices in place, delegation still carries some risks. You can never control everything, especially when it comes to people. Tasks may get deprioritized, misunderstandings can happen, or someone might not deliver to the level you expected. You might realize too late that the person didn't feel comfortable asking for help. These things happen.

So, even though these steps aren't a guarantee that everything will go perfectly, they significantly reduce the chances of things going wrong. And even when things do go wrong, you can recover. What's far riskier is trying to do it all yourself: that's the path to burnout, bottlenecks, and a team that never grows.

Mastering delegation means being intentional at every stage. From deciding what to delegate and to whom, to adjusting your style based on the situation, setting clear expectations, and following through with accountability, each step plays a role in making delegation really work.

Why Tech Leads Struggle to Delegate and How to Overcome It

Some tech leads feel constantly overwhelmed, juggling too many things at once: coding, one-on-ones, alignment conversations, and putting out fires, all while trying to keep up with strategic initiatives. And despite working long hours and being in endless meetings, there are always more things they know are important but never get around to. They feel like their days slip away, running from one task to another, without ever having time to step back and plan.

Yet, even when it's obvious they need to offload work, they still hesitate to delegate. Why? Because deep down, they're afraid: afraid that the outcome won't be as good as if they did it themselves, afraid of not getting credit, afraid of the effort required to teach someone else, and ultimately, afraid of losing control.

The reality is that the main reason tech leads start delegating isn't because they want to; it's because they have to. Eventually, it just becomes impossible not to.

This struggle is completely normal, especially for tech leads transitioning from an individual contributor role. As ICs, they were used to full ownership: knowing every detail of their work, delivering high-quality outcomes, and receiving direct recognition for it. Now, suddenly, success is no longer about what they personally deliver but about what their team delivers. That shift is not easy, but there is no way around it, as great leadership is about scaling impact.

In this section, I'll dive deep into the most common fears tech leads have around delegation, exploring practical ways to overcome them. The goal is to help you ease into delegation with more confidence, so you can reclaim your time, grow your team (and yourself), and ultimately improve overall team performance.

AFRAID OF LOSING CONTROL

Delegation means enabling others to do the work, which is often harder than just doing it yourself because it requires you to let go of control.

Most tech leads struggle with this, especially in the beginning. They're used to having full ownership over their own work as individual contributors in the team, knowing every detail, and being in control of outcomes. When they step into a leadership role, they try to maintain that same level of control across the entire team. This often leads to micromanaging: reviewing every PR, being part of every decision, attending every team conversation. Many tech leads have tried this, but it's just not sustainable. Eventually, they burn out.

And the worst part is that no matter how much effort you put into controlling everything, you're never truly in control. There are too many moving parts, too many variables that can shift unexpectedly. Trying to control everything creates frustration, both for you and your team. You feel like things are slipping through the cracks despite your best efforts, and in response, you try to control even more, trapping yourself in a cycle. The only way to break free from this loop is to accept that you can't control every little detail. What you can do is prepare for uncertainty and learn how to navigate things when they don't go as planned. That's what experienced tech leads continuously try to do.

At the root of most delegation struggles, whether it's worrying about the outcome, thinking "it's quicker and easier to just do it myself," or even fearing that you won't get the credit, lies the fear of losing control. It's the biggest mental challenge tech leads face when letting go.

This fear is especially strong because delegation requires vulnerability. It means trusting others with responsibilities you used to own. In leadership, vulnerability is often misunderstood as a weakness, when in fact, it's one of the greatest strengths of a successful tech lead, and a powerful way to build trust.

If you want your team to trust you, you need to trust them. Struggling to let go is often a sign of distrust: worrying that they will drop the ball, that they won't do the task as well as you, that things won't turn out exactly how you want. And if you don't know what your team is capable of, that fear makes sense. But the only way to find out is to take chances on people, give them opportunities to prove themselves. You might be surprised at what they can do when given the space to own a task.

So, I invite you to try. The best way to start overcoming the fear of losing control is to practice delegation in small steps. You don't have to hand off everything at once, and there are ways to do so without micromanaging. The secret is slow exposure: start with a low-risk task, set clear expectations, and check in without hovering. By applying the strategies from the section "Delegation Process: Step-by-Step" on page 143, setting clear expectations, having a transparent tracking process, and understanding your role in delegation, you can maintain visibility without needing to be involved in every step.

A simple task to start with is asking someone else to facilitate the daily standup. This is a common action that tech leads, focused on control, do to ensure they are getting all the necessary updates and information to stay up-to-date with everything that is happening, often saying, "No one else will do it if I don't." But the opposite can be true too: no one else jumps in because they know you are always the one doing it. So, try telling your team that you'd like to mix it up a bit, and ask who would like to take over or come up with a plan for a rotation process. This might be an easy way to start exploring delegation since you're still part of those conversations. You can continue asking the questions you need to ask and make sure you get the information, even if you're not the one facilitating. Plus, the chances of things going completely sideways are smaller because they've seen you run it for a while now.

Once you agree on a process for facilitation, set some ground rules. For example: we go over the board left to right (Backlog → To Do → In Progress → In Review → Testing → Done), discussing progress and potential blockers with the goal of going through all the tasks within the reserved time, leaving a few minutes at the end for announcements like "I have a medical appointment from 3 to 4, so I will not be available." Say the things that sound obvious; those are the ones that generate the most misunderstanding and conflicts in the team. Also, keep in mind that this is a continuous process. If you feel something isn't being addressed correctly, you can always discuss it with your team or give individual feedback.

The same applies when handling urgent bugs reported by clients. Instead of taking the lead and verifying what is happening yourself, agree on a process with your team to handle them together, perhaps by rotating people, for example. To help you feel more in control at first, you can ask them to escalate to you if there's a bigger problem or if they can't handle it within a certain time threshold.

Sometimes, losing control is exactly what needs to happen. When you give people ownership, they bring fresh ideas and approaches you may not have considered. It will also help you develop resilience and learn how to deal with things when they go wrong.

I can say from experience that once you experience the benefits of letting go of control, you won't want to go back. I started off as a micromanager and burned out one month into the role. But I learned to do better, and delegation was a game-changer.

Over time, you'll see that letting go doesn't mean losing control. It means gaining a stronger, more capable team. By embracing vulnerability and letting go of control, you'll scale your positive impact as a leader in tech: you'll create growth opportunities for your team members, and both they and you will learn and grow together.

WORRIED ABOUT THE OUTCOME

The second common fear tech leads have when delegating is worrying about the outcome of the delegated task not being as good as if they did it themselves.

This can be addressed by setting clear expectations.

I remember when I was talking to another tech lead about a situation he had gone through. He had delegated the task of writing technical documentation for a new feature to one of his team members. He had originally planned to do it himself but, like many of us, got overwhelmed with other tasks. So, he decided to ask a team member to handle it instead.

He gave the person a vague instruction: "Please write up the documentation for the new feature so the team can understand how it works." He didn't specify things like the format, the level of detail, or even if things like code examples, diagrams, or edge cases were necessary. He assumed that the team member knew what was expected, based on the way other technical documents had looked before.

A day later than expected, the documentation came in. But when he reviewed it, he was disappointed. It was just a brief overview, explaining what the feature did, but it lacked real depth. There were no architectural diagrams, no edge case handling, and no detailed explanations that would have made it helpful for the team. It was functional, but far from what he needed. In the end, he ended up spending more time updating and fixing the documentation than he would have if he had just written it himself from scratch.

He shared with me how frustrated he was, but as we discussed it, he realized he was partly to blame. He hadn't set clear expectations from the start. He hadn't explicitly stated what level of detail was needed, who the documentation was for (developers, QA, or nontechnical people), or even the key components like code examples, error handling, or diagrams.

When he reflected on it and talked about it with another colleague, he came to the conclusion that, had he set proper expectations, this could have been avoided. The next time he delegated a similar task, he was much more specific about what he expected, and the outcome was completely different.

This is a common story I hear from tech leads, and it's a great example of how important it is to be clear and specific when delegating tasks, especially something like documentation, where the format, level of detail, and intended audience can really affect how useful the end result is. (More on this in the section "Set Clear Expectations" on page 151.)

Giving very specific instructions is a way to ease into the process of delegation and get more comfortable with it. But it shouldn't always be your default strategy. As you get more into the habit of delegation and gain more trust in your team members, your guidelines should become more high-level. Focus on what the outcome should be, share the information you already have on the task, agree on a timeline, and leave the rest to them. Give them more autonomy to figure out how to address the task, who to talk to, and how to approach the solution. You might be surprised by how people can positively surprise you when given the chance.

When you give others the space to add their own touch to a task, they often come up with better solutions than you might have. They have a more objective view of the problem or maybe a different experience to draw from. A lot of great things can come from letting people put their own spin on the task. They're more motivated because they can get creative, instead of just following a list of steps, and the solution they come up with could be even better than yours, ultimately benefiting the entire team. And keep in mind that whatever good work your team produces reflects positively on you as a tech lead. This has happened to me.

There was a time when I delegated a complex task of optimizing a performance bottleneck in our application. I had initially thought of a solution that I thought would be the best approach, but one of the developers on my team came up with an alternative method. She suggested a faster, more efficient solution that was more streamlined than my own idea. After reviewing it, I realized that her approach was not only faster but also easier to implement with fewer changes to the existing codebase. The result was a performance boost that exceeded our expectations in both speed and stability, and she felt more empowered because she was able to take ownership of the solution. It made me realize that sometimes stepping back and trusting the team can lead to even better outcomes than I had initially anticipated.

This brings me to the point that while delegation starts with providing clear expectations, it also grows into empowering your team. Trusting them with autonomy can lead to great things.

"IT'S QUICKER AND EASIER TO JUST DO IT MYSELF"

As a tech lead, there's always that one task you've been doing for what feels like forever because you know how to do it best. Maybe it's debugging the CI/CD pipeline when it fails, because you've been in the team the longest, were there when it was first set up, and know all its moving parts. You've done it a dozen times, it doesn't break often, and fixing it usually takes only 15 minutes. It's easier to just take care of it yourself rather than delegate it to someone else. It's a scenario most of us have been in: feeling like it'll just take too long to explain to someone else, and you can knock it out in half the time.

But that 15-minute fix comes at a cost. Every time something breaks, it derails your day. Beyond the disruption, context switching, and handling something that could easily be done by someone else in the team, you are also putting your team at risk. If the issue happens while you're on vacation, out sick, or just unavailable, your team is blocked from deploying. They're stuck, unsure where to start, and delivery stalls. Also, by always being the one to jump in and fix it, you're creating a knowledge silo; no one else gets the chance to learn the process or improve upon it.

Now, let's say you pass the task on to someone else. Here's how you could approach it.

During your next standup, say something like "I've been handling CI/CD issues when they pop up, but I'd like to teach someone else how to manage them. Who's interested?" You'll likely get a volunteer who's eager to grow in that area.

Alternatively, bring it up in a one-on-one. If you notice someone who's curious or looking for a new challenge, you might say: "I usually take care of this, but I think it's a great opportunity for you to learn. Want to pair on it this time?" That way, you're not just delegating a task; you're giving them space to grow and build confidence in a new skill.

Imagine next week, the issue comes up again. Instead of jumping in solo, you spend 30 minutes pairing with them to fix it, or better yet, record a walkthrough or help them document the process. Maybe you'll spend another 30 minutes answering follow-up questions. But once that knowledge is shared, the task is no longer solely yours to manage. You might even encourage them to share what they've learned with someone else in the team, spreading the knowledge even further. Over time, your team becomes capable of handling it independently. It might take a little effort up front, but soon enough, you'll no longer have to worry about it. The delivery risk will be removed and your team will be stronger for it.

Delegating tasks like this may take effort up front, but the long-term payoff is clear: your team becomes more capable, confident, and independent. You're investing in your team's resilience. Ultimately, that's the job of a great tech lead: scaling both your impact and the success of your team.

AFRAID OF LOSING THE CREDIT

This fear is common but often unspoken. Many tech leads hesitate to delegate because they worry that if they're not the one directly producing the work, their contributions will go unnoticed. They fear that their impact will be invisible, that leadership or peers will recognize only the person who completed the task, not the one who made it happen.

I want to start by saying that this is a very normal fear and nothing to be ashamed of. The reason is simple: as individual contributors, tech leads were used to full ownership of their work, the outcomes, and all the praise that comes with it. This is how they've grown, how they've been evaluated, how they've built their careers. Promotions and recognition have always been tied to what they did, not necessarily what they enabled others to do. Shifting away from this mindset is complicated and takes time.

If this fear resonates with you, here's a story that might feel familiar. A tech lead I know was working on a critical performance optimization for their product. The issue had been on the radar for months, but due to other priorities, she hadn't been able to focus on it. When the time finally came to address it, she had a clear idea of how to tackle the problem but decided to delegate the task to a senior developer on the team so that it wouldn't be delayed any longer.

She outlined the approach, provided guidance on what to look for, and even suggested a few potential solutions based on her prior research. She checked in regularly, helped troubleshoot blockers, and refined the approach when necessary but made a conscious effort to let the developer take ownership.

A few weeks later, the developer successfully implemented the optimization, and the results were impressive: query response times improved by 40%, and system stability significantly improved. When the work was presented in a company-wide demo, leadership and peers applauded the developer for his great work. The team was also recognized for solving a long-standing problem, and the developer even got a personal shoutout from leadership in an all-hands meeting.

The tech lead, however, received no direct recognition. The developer personally thanked her, but she would have liked others to know how she had guided, mentored, and provided solutions behind the scenes. While she was genuinely happy for the developer, she couldn't shake the feeling of being invisible in the process.

This situation is exactly why many tech leads struggle with delegation. It's hard to let go when you fear that your impact will go unnoticed.

But after some time, she started to see things differently. In her following one-on-ones with stakeholders and managers, she heard things like "That was a great achievement your team delivered. Sounds like you're doing a great job leading them and prioritizing tasks like this" or "I heard you had something to do with it."

She realized that the credit did come back to her, just in a different way than she was used to. And when it came time for performance reviews, she wasn't writing, "I improved system performance by X%" but rather, "Led an initiative that improved system performance by X%." That subtle shift in framing made all the difference.

As she continued to delegate more, leadership started recognizing her not just as an individual contributor but as someone who was building and enabling a strong, capable team. That's when it clicked: true leadership isn't about individual credit; it's about scaling impact. The fact that her team could execute at a high level without her needing to do everything herself was the real win.

As a tech lead, your success is no longer measured by the number of tasks you complete. Your value isn't tied to how much you deliver personally but how much your team delivers. That's what your stakeholders and leadership care about.

A big part of transitioning to leadership is understanding that it's not about you anymore. Your results are now your team's results. This shift requires moving from an I mindset to a We mindset:

- Instead of doing it yourself, empower your team to take ownership.

- Instead of controlling every detail, focus on making your team more autonomous.

- Instead of optimizing for the best short-term result, optimize for sustainable long-term growth.

- Instead of constantly checking up on or redoing your team's work, trust them and provide guidance when needed.

- And most importantly, instead of just taking accountability when things go wrong, share the credit when things go right.

In the situation mentioned earlier, the tech lead was very involved in the success of that performance optimization. But the best part is that, as a tech lead, you also get credit for things you weren't even directly involved in. Anything your team delivers is a reflection on you. Whether you were closely involved in the task or not, you will always be associated with the results of your team. The credit might not be instant or as direct as when you were an individual contributor, but it's there. And more importantly, the impact of your work is multiplied by the people you lead.

Key Takeaway

Stepping into a tech lead role means redefining how you create impact. Your value isn't tied to doing everything yourself; it comes from how well you guide and support your team to succeed.

Delegation plays a central role in that shift. When you delegate effectively, you create space for your team to grow, take ownership, and develop confidence. At the same time, you gain the bandwidth to focus on higher-level priorities like team alignment, system design, and long-term strategy.

This shift can feel uncomfortable. It takes time to build trust in others and in your own ability to lead without micromanaging. You may feel the urge to stay involved in every detail, especially if you're used to being the go-to person for execution.

But strong leadership means creating an environment where great work happens through others. When you invest in delegation, you're investing in your team's strength and in your own evolution as a leader. That's what unlocks real, sustainable impact.

7

Building and Scaling Tech Teams

Besides building relationships, running one-on-ones, providing feedback, and growing people through delegation, which are the foundational responsibilities of a tech lead to ensure smooth day-to-day team operations, you'll also take on broader team-building responsibilities. These include recruiting new team members, onboarding them effectively, and conducting performance reviews, all while shaping the team's culture and supporting its growth.

In this chapter, scaling doesn't just refer to growing your team in size; it's also about taking your team to the next level in terms of capability, structure, and effectiveness.

I'll share practical guidance on these aspects of the role, drawn from my own experience. You might face these responsibilities sooner than you expect, and being unprepared can slow down your team's momentum or lead to avoidable friction.

I'll also cover three key challenges that often come up when building and scaling tech teams: building an onboarding process from scratch, enabling effective collaboration, and addressing underperformance when it arises.

These challenges are complex, but they're also solvable, with the right mindset, tools, and systems in place. This chapter aims to give you a starting point to tackle them confidently and proactively as your team grows.

Recruiting and Onboarding Developers

The amount of effort you'll need to invest in recruiting and onboarding depends on factors such as the size of your team and the reason for the opening, whether you're building a new team from scratch, scaling an existing one, or backfilling for someone who left. While these tasks might not be part of your day-to-day work, they become a key focus when the need arises.

In more established teams, recruitment may come in waves, triggered by new projects, team members leaving, or evolving business needs. Sometimes, the hiring need isn't about growth but about balance, bringing in the right mix of experience levels, skills, and interests to ensure strong team dynamics, well-rounded coverage, and no one being stretched too thin.

As a tech lead, your level of involvement in hiring depends greatly on your company's size and structure. In some organizations, tech leads are expected to co-own the process alongside engineering managers or internal recruiters. In others, you might simply provide input at key stages, such as designing the technical evaluation or participating in final decision making.

Even when you're not leading the process end to end, your input remains critical. You're often best positioned to assess a candidate's technical fit for your team's specific work, how well they'll mesh with the team's culture and ways of working, and whether they bring complementary skills or perspectives.

You may also be asked to help shape or adapt existing hiring processes by refining interview rubrics, identifying gaps in the evaluation loop, or helping to clarify what "good" looks like for a particular role. And while you might not be the one pushing the process forward alone, you'll still need to ensure it aligns with your team's needs and timelines, and that whoever joins is set up to succeed from day one.

This section will help you step confidently into this responsibility, covering what to focus on when interviewing and how to design an onboarding process that supports both the new hire and the team as a whole.

IDENTIFYING TEAM NEEDS

A balanced team is one that meets the needs of the organization while staying effective and growing sustainably. It's about having a good mix of experience levels: junior developers, mid-level developers, senior developers, and leads (tech leads, staff engineers, principal engineers) so that responsibilities are covered without gaps or overload.

When a team lacks this balance, problems start to appear.

Many businesses assume that a team made up entirely of senior engineers and leads will run smoothly with fewer issues. In reality, that setup brings its own challenges. With too much seniority, responsibilities can start to overlap, leading to inefficiencies and, at times, conflicts over ownership. For example, when multiple senior engineers all want to take the lead on a new initiative, it can create competition and friction instead of clear direction. Without clear role distribution, decision making can become slow and contentious, impacting overall team productivity. Everyone can do everything, and everyone is expected to do the same things.

This can go in two directions. One possibility is long conversations and debates whenever a decision needs to be made, as everyone has experience and a strong opinion on how things should be; seniors tend to be particularly opinionated. The other outcome is that some people become unmotivated and hold back their input because they feel there isn't enough space for their voice to be heard.

Also, everyone tends to aim for the same career advancements. Seniors often aspire to move into staff engineer or principal roles, but there are usually limited opportunities for these career advancements. This increased competition can lead to frustration and stagnation, making some individuals feel stuck or demotivated. This means more competition and less opportunities for upscaling. This can quickly demotivate people, causing them to become complacent, stop pushing themselves, or, in some cases, burn out or leave altogether.

There's also the issue of pressure: when everyone is senior, some may hesitate to ask questions they consider basic or challenge their peers, fearing they "should already know this."

Junior engineers, on the other hand, bring fresh perspectives. They ask questions that challenge assumptions and encourage rethinking long-standing practices because they want to learn. They might also see activities like cleaning feature toggles as interesting and as an opportunity to learn something new, which will make them prioritize them, making it a win for the whole team, whereas seniors might find it boring. A team without juniors might lose out on this energy and innovation.

Another downside to having a team of only very experienced members is the lack of mentorship opportunities. If there are no less experienced devs to mentor, senior engineers miss out on developing leadership skills. I once saw a team where all the engineers were senior-level really struggling to make decisions. While technically strong, they often got stuck debating architectural decisions for too long. When they brought in junior developers, explaining concepts to them forced the seniors to clarify their thinking, leading to faster and better decision making.

On the other end of the spectrum, a team made up entirely of less experienced engineers faces different struggles. Without experienced team members to learn from, they rely on trial and error, which slows down progress and increases mistakes. This also puts a huge burden on the tech lead, who becomes the only source of guidance. Without additional experienced engineers to help set standards, code quality and architecture will suffer.

I know a startup that decided to hire only junior engineers to cut costs. While they were enthusiastic and quick learners, they struggled with structuring their codebase efficiently. Without experienced engineers to guide them, they ended up with an overly complex system filled with redundant logic, making maintenance and scaling the business difficult. Eventually, the company had to bring in senior engineers to refactor large parts of the code, costing more time and resources than if they had built a balanced team from the start.

To build a truly balanced team, you need to think beyond just roles and titles. A balanced team means having a healthy mix of experience levels, skill sets, interests, and motivations. This kind of diversity brings many advantages: broader coverage of your product's needs and greater innovation through varied perspectives.

For example, in a full-stack team, some people might prefer frontend work while others are more interested in infrastructure. This balance ensures that different areas of the system get proper attention. This does not mean creating knowledge silos; it's just about having people who will go the extra mile when it comes to that particular topic because they are more interested in it.

But achieving this balance requires intentionality. Start by asking yourself: What problem am I trying to solve? Are there skill gaps on the team? The good news is that answering these questions and deciding how to move forward is a shared responsibility. You'll work closely with your manager, product managers, and other key stakeholders to make these decisions together and ensure the team is set up to deliver effectively.

If the issue is experience, consider whether some of your junior engineers are ready to take on more responsibility. Promoting from within can inject new energy into the team while solving capability gaps. If the issue is a lack of specific technical skills—perhaps your team is about to adopt a new technology that no one is familiar with—this might require hiring externally or temporarily bringing in experts from other teams to share knowledge.

Don't just default to thinking, "I need two more seniors." Instead, reframe the question: "What outcomes are we missing because of the current imbalance?" Maybe the team is too quiet and lacks fresh perspectives. In that case, bringing in enthusiastic junior talent can help. Or maybe the team debates endlessly without clear decisions, pointing to a need for stronger leadership or clarity.

Also, keep in mind that every team is different. They're at different stages of growth and have unique challenges. When joining or starting a new team, the key is to assess where they are right now and what's holding them back. Once you understand that, you can decide whether the answer is promotion, hiring, restructuring, or something else.

The best part of a well-balanced team is that it creates a strong learning culture. Juniors have mentors to guide them. Seniors and leads get leadership opportunities. The tech lead can focus on strategy instead of being the sole source of knowledge. And with diverse perspectives in the mix, the team becomes better at solving problems and thinking critically about its work.

EFFECTIVE RECRUITING AND INTERVIEWING STRATEGIES

At some point as a tech lead, you'll find yourself hiring people for your team. The level of your involvement may vary depending on the company, but you should at least be aware of the process and know what to expect. Here are some strategies that can help you recruit and interview tech candidates more effectively:

Get clear on why you're hiring

> You might think, "Obviously, we need more people." But take a moment to get specific. Are you hiring because someone left and there's a team size quota? Because you're starting a new initiative and need to build a team from scratch? Because you're overloaded with work and the current team can't keep up? Or maybe you're missing a key skill set or trying to rebalance your team across experience levels? The answer to this question shapes everything: how you interview, what you look for, and how the new hire will integrate. I've seen engineers struggle when they join a team with vague expectations. And when that's the case, guess who they turn to for clarity? You. That lack of alignment not only frustrates them but slows everyone else down.
>
> When you're clear on why you're hiring, it becomes much easier to answer that common interview question: "How will I contribute to the team?"

Get clarity on the whole process

> Even if you're not involved in every step of the hiring process, it's incredibly useful to understand the full journey a candidate goes through. Before jumping into interviews, take some time to familiarize yourself with the

entire flow. Look for any documented processes and have a check-in with the HR person responsible for the process and work closely with this person to understand what each stage looks like, where you fit in and what's expected of you, how you will track candidates and feedback, and how you will collaborate during the process. To get even more context, jump into a quick conversation with another tech lead who's been part of recent hiring rounds.

This broader view will help you prepare for candidate questions, and the close collaboration with HR will make the process smoother on all sides but also might trigger some ideas for improvement of the overall process.

Write a job description that reflects reality

Sometimes you get a boilerplate job description handed down from HR or a company-wide template. If you are interviewing for your team and can influence or adjust it, absolutely do it. Add context that's specific to your team. What problem are you solving? What stage is your team at? What kind of collaboration do you expect: do you pair regularly, do you work async, are there on-call rotations? What are the must-have technical skills and the nice-to-haves? When candidates understand the real picture, you attract people who are both more qualified and more enthusiastic about your specific context, not just "a backend role somewhere."

Even if you're conducting interviews that aren't directly for your team, like contributing to a company-wide hiring pipeline or helping out another team, don't overlook opportunities to improve the process. If you notice something that could be better, share your feedback with HR or the hiring coordinator. Whether it's a gap in the interview loop, unclear evaluation criteria, or a misalignment between job descriptions and reality, your input can make a big difference. Improving the hiring process benefits not just your team but the entire organization.

Make the process inclusive (on purpose)

Inclusion doesn't start in the interview room. It begins with the job description and runs through every step of the hiring process.

Even small details, like how you write a job ad, can influence who applies. Certain phrases or tone choices can unintentionally discourage women or underrepresented candidates from applying.

Interview format matters too. A two-hour whiteboard session in front of multiple interviewers might feel standard to some but intimidating or exclusionary to others. That's why flexibility matters.

Ideally, your process offers a few different options and adapts to the needs of individual candidates. For example, someone with caregiving responsibilities might prefer the flexibility of a take-home task. Take-home tests, pair programming sessions, and async exercises all have their pros and cons, and no one format works for everyone.

That said, adapting the process to each candidate takes coordination and effort. It requires alignment across the hiring team and more flexibility in scheduling, which is why only a handful of tech companies have been experimenting with this kind of customization so far.

But even if you can't offer this level of flexibility just yet, it's worth rethinking your current approach. Instead of sticking with "This is how we've always done it," start small. Test changes. Pay attention to the candidate experience.

Also, you can only hire a diverse team if you're starting with one. Homogeneous teams tend to hire in their own image, even unintentionally. If your interview panel lacks diversity, try borrowing interviewers from other teams or departments, especially those who can bring different perspectives. You can also introduce external tools to help reduce implicit bias in your filtering and evaluation process. Injecting diverse voices into hiring decisions at every stage makes it far more likely you'll build a team that reflects a broader range of experiences and ideas.

Your hiring process should aim to get the best out of each candidate by creating a fair, inclusive environment that mirrors the actual work of the role, without introducing unnecessary barriers.

Keep an open mind when interviewing

We all carry biases, even if we don't realize it. It's very natural to gravitate toward people who share your background, your interests, your career path. But that mindset can really narrow your field of view. You risk overlooking someone who could bring fresh ideas, different experiences, or complementary skills to your team. Not only that, but you slow down the process because your bar becomes too specific.

The best teams aren't made up of clones; they're made up of people who cover each other's gaps. So when you're interviewing, pause and reflect on what you're actually evaluating. Are you looking for familiarity, or for someone who could challenge your assumptions in a good way?

Involve the whole team

Hiring shouldn't rest solely on your shoulders. The team will work with the new person, so they should share responsibility in choosing them. That shared ownership often leads to better hires and a stronger team culture overall.

Involving the team doesn't mean the candidate needs to meet every single person during interviews. But everyone should have a chance to contribute to the process. That might mean helping draft the job description, participating in technical interviews, or leading culture interviews. A common trap is letting only the tech lead or senior engineers handle interviews, which misses out on a bunch of benefits. When more people are involved, you can run interviews faster and avoid bottlenecks.

It's also a growth opportunity. Interviewing builds soft skills like asking thoughtful questions, really listening, and clearly explaining complex ideas. Plus, more voices mean a better chance of spotting whether someone is a good culture fit. And from the candidate's side, it makes the process feel more authentic, as they're not just meeting the leadership; they're meeting the team they might actually work with.

Constantly improve as an interviewer

Interviewing is a skill like any other. Just because you're expected to jump into it doesn't mean you're instantly good at it. And that's totally OK. Start small. Reflect after each interview: Did the conversation flow naturally? Did you find yourself repeating questions or running out of time? Were there moments where you weren't sure what to ask next? Talk to someone with more interviewing experience and bounce your challenges off them. Ask how they would approach the situations you found tricky.

If you're pairing during interviews, like we used to do at Thoughtworks, take advantage of that setup. Ask your partner for feedback. Did they notice anything you could do differently next time?

And don't forget the most direct source of feedback: the candidate. While this might be new for your company, or not part of the default process, it's worth exploring. Sometimes it requires building a stronger relationship with HR or recruiting so that feedback isn't filtered or lost. Hearing how the interview experience felt from the candidate's side can uncover blind spots and help you improve faster.

Improving your interview skills is good for your team and for you. The more prepared and confident you become, the less stressful and more effective interviews will feel.

Get involved in sourcing candidates

It's easy for tech leads to take a backseat when it comes to sourcing, assuming it's entirely the recruiting team's job. But you can have a real impact here, and the earlier you get involved, the more influence you have over the type of candidates coming through the pipeline.

Start by thinking about your own network. Are there people you've worked with before or know from the community who might be a good fit? Reach out. Be vocal about hiring when you attend meetups or industry events. A simple post on LinkedIn sharing what your team is working on can go a long way.

You can also help by reviewing inbound applications more intentionally, flagging promising candidates early on, or sharing feedback to refine the sourcing criteria.

The more you engage, the better the odds you'll attract people who are genuinely excited about your team and the problems you're solving.

The more you invest in the hiring process, from sourcing to interviews to team involvement, the more you reduce the risk of hiring the wrong fit and having to go through the whole thing again. Being intentional up front saves you a lot of time, energy, and disruption down the line.

ONBOARDING EFFECTIVELY

A smooth onboarding process is a key part of a high-performing team and will have a great impact on your overall team success. A well-structured onboarding process enables new hires to ramp up quickly, work autonomously, and feel like valuable contributors from day one. It ensures they gain mastery over their daily tasks and develop a solid understanding of the team's history, expectations, and ways of working. Yet, it's often treated as an afterthought in tech teams, leading to unnecessary confusion and frustration.

I've seen people leave companies because of a poor onboarding experience. A staff engineer once joined a large tech company only to find no clear documentation, no connections between systems, and no defined expectations for her role, leaving her frustrated and unsupported. Six months later, she left. And she's not alone. Many companies underestimate how critical those first few months are for retention and productivity.

Given the fast-changing nature of the tech industry, ensuring new joiners ramp up quickly and contribute effectively has never been more important. A great onboarding process means guiding new hires through a structured, engaging experience that helps them integrate smoothly, rather than simply overwhelming them with documentation.

The best onboarding experiences share common characteristics. They are intentional, structured, and involve the whole team. In the following sections, we'll go over what they look like in practice.

Start onboarding before day one

Onboarding begins the moment an offer is accepted. A simple welcome email with practical details, like what to expect on the first day, an agenda, and key contacts, sets the tone. For remote employees, laptops and other equipment should arrive before their start date. For in-office hires, everything from desk setups to system access should be ready. Account setups should be completed in advance, or at the very least, there should be a clear process for requesting access to essential tools, so new hires aren't stuck waiting during their first few days.

While these tasks may technically fall under other departments, it's still valuable for tech leads to be aware of the pre-day-one experience. It directly impacts how a new hire shows up on their first day: whether they feel confident, welcomed, and ready to contribute or confused and disconnected.

If you notice gaps in this early process, don't hesitate to offer feedback or collaborate with the relevant teams to improve it. Even small improvements here can make a big difference in how smoothly onboarding begins. One simple but effective step you can own: keep a checklist of the essential tools, systems, and services your new team member will need in their first weeks. This gives you a quick way to double-check readiness and, if something is missing, helps you guide the new hire to the right people who can unblock them.

Involve the whole team

As a tech lead, you don't have to do all of this yourself; you just need to make sure it gets done.

Too often, I see tech leads taking on the full burden of onboarding, preparing sessions, answering every question, and trying to manage it all alone. Not only does this overwhelm you, but it also robs your team of mentoring opportunities and fresh perspectives on improving the onboarding process. When only one person owns onboarding, it reflects a single viewpoint, and we all know that can be limiting. It's a great opportunity to delegate (you can find the whole delegation process in Chapter 6).

A common approach is assigning a buddy to the new joiner from day one until onboarding is complete. A buddy is typically a peer, often on the same team, making the process more personal and helping to fill in any gaps. They take ownership of onboarding tasks, plan sessions, and provide day-to-day guidance. Being an onboarding buddy gives new hires the support they need, and at the same time, it offers engineers a valuable opportunity to grow professionally.

While a buddy can deliver all the onboarding sessions, I prefer involving the whole team. In my teams, we split onboarding topics among different members: high-level product overview, architecture, path to production, coding environment setup, ways of working, and team values. Each person prepared documentation, shared it with the team for feedback, and updated it accordingly before presenting it to the new joiner. Even the product owner was part of the process. This not only helped the new joiner meet the team and start building relationships but also encouraged shared ownership of onboarding.

As a tech lead, make sure the buddy has time to prepare and support onboarding. Expect that their usual workload will be reduced for the first few weeks as they focus on helping the new joiner settle in.

Another way to involve the team in the onboarding process is to have them pair on tasks with the new joiner. Pairing is one of the fastest ways to get a new joiner up to speed and integrated into the team. These sessions naturally become mini knowledge exchanges, as they help the new joiner explore the codebase, understand past decisions, and learn any quirks in the tooling. The best part is that no extra preparation is needed, just a task to work on and time to collaborate.

Define a clear plan with actionable steps

A well-structured onboarding process includes a clear plan with actionable steps to help new joiners ramp up quickly. In the first few days, they should get an overview of essential topics, from how the company operates to the tools they'll use to ship code.

A good onboarding schedule includes several key sessions:

Company values and mission
> Understanding the broader vision

Team ways of working
> How the team collaborates and communicates

Product definition
> The team's role within the company and its impact

What the team is building and why
> The purpose and goals behind their work

Many teams skip these high-level sessions and jump straight into technical details, but this foundational knowledge is crucial. Once the big picture is covered, onboarding should move into a few key areas:

Architecture overview
> How the system is structured

Path to production
> Steps to deploy code successfully

Pairing sessions
> Hands-on guidance for setting up the development environment

To support these meetings, provide documentation that new hires can reference later, including the following:

Architecture diagrams
> A visual representation of the system

Path to production guide
> A step-by-step breakdown of deployments

Product documentation
> Key details about what the team is building

Tooling overview
> The technologies and platforms used

Setup instructions
> How to configure their local environment and access essential tools

This documentation can quickly become outdated. Maintaining accurate, up-to-date docs requires discipline, but when it's treated as a shared team responsibility, the effort becomes much more manageable. And this is one of the few types of documentation truly worth investing in keeping up-to-date, as it has a direct impact on how quickly and smoothly new team members can get up to speed.

Having a structured onboarding guide ensures new joiners have all the information they need to become productive quickly.

Track progress

Onboarding is a team effort, and tracking progress should reflect that. My approach is to create a task on the main team board, assigning both the new joiner and their buddy as owners. This makes onboarding visible and ensures shared responsibility. After all, it affects the whole team and our delivery; even if a buddy has been assigned, it's not just one person's job.

By having this task on the board, the new joiner is included in the team's workflow from day one. They have a task to track, give updates on, and, most importantly, use as a way to raise blockers in standups. As we all know, it can be intimidating for a new hire to speak up and ask for help. This setup makes it easier and encourages them to engage early.

Another benefit is that it immediately gives them a sense of contribution. Talking about onboarding in stand-ups reinforces that they're part of the team and helps them build confidence as they navigate their first few weeks.

Continuously improve

You can't improve what you don't measure. Without tracking the impact of changes, you won't know if your onboarding process is actually getting better or making things worse.

Some teams track onboarding efficiency by measuring how quickly a new hire starts contributing to the codebase. For instance, getting a pull request merged in the first week can signal a few positive things: a solid CI/CD system with guardrails, a culture of frequent iteration, a fast-moving organization, and a practical onboarding process that gets people working quickly.

In fact, some teams go as far as encouraging new hires to ship code to production on their very first day. And while this is a popular idea in tech circles, I'm personally not a big fan.

Shipping something on day one can demonstrate maturity in your systems and processes, but it's not always the best experience for the person doing it. For some, it's an exciting challenge. For others, it creates pressure and sends the wrong message, like that speed is more important than integrating in the team or understanding the context.

There are easier, lower-pressure ways to introduce someone to the codebase, like pairing or helping them set up their environment.

One practice I often encourage is giving new joiners full access to update onboarding documentation, especially setup guides. These are typically the first things to become outdated as tools evolve. Some teams hesitate to allow this, worrying that someone without full context might make incorrect edits. But who better to spot unclear steps than someone going through the process for the first time? It's a chance to make the process clearer and give them an immediate sense of ownership and involvement. If you're concerned about accuracy, simply ask their onboarding buddy or a more experienced team member to review changes before they're finalized.

Of course, just because someone pushes code doesn't mean they're fully onboarded. They still need to meet the team, understand the product and its purpose, get to know the system, and learn how the team works.

That's why I prefer other metrics for tracking onboarding efficiency: ones that reflect the full experience.

For example, you can track the time it takes to complete an onboarding checklist: attending onboarding sessions, gaining access, reading the starting documentation, and setting up the development environment.

Others use lightweight surveys or informal conversations to gather feedback. Was the experience smooth or frustrating? Did they feel supported? Whether or not this is your primary metric, I highly recommend building in feedback loops. Ask new hires to note what felt confusing or missing while the experience is still fresh.

The key takeaway: there's no one-size-fits-all metric. What matters is that you're tracking something. Whether it's checklist progress, feedback, or something else, choose a signal that fits your team and use it to learn and improve.

Because the real issue is that many teams don't track onboarding at all. They simply assume "it'll get done when it's done," leaving them with no way to estimate its impact on the team, delivery timelines, or overall efficiency.

The goal is to recognize onboarding as a system worth improving, then gather feedback, measure progress, and iterate on it continuously.

Building a High-Performing Team

Having, or better said, leading a high-performing team is the ultimate proof of a great tech lead. It's every tech lead's dream to have a team that enjoys working together, feels motivated about what they're building, and runs like a well-oiled machine. A team that's seen as reliable, efficient, and capable of consistently delivering impact.

But it's also one of the hardest things to achieve.

Most teams don't struggle because they lack talent or tools. They struggle because they have the wrong idea of what a high-performing team actually is, and more importantly, what behaviors, principles, and habits are needed to build one. That's why in this section, I want to break some of the most common myths around high-performing teams and instead give you a clear, practical view of what "high-performing" really means and how to get there.

I'll start by redefining what a high-performing team truly looks like, beyond just fast delivery or coding velocity. Then, I'll walk you through the five key dynamics that make these teams successful, how to build and protect a healthy team culture, how to create and maintain psychological safety, and finally, how to evaluate if your team is actually high-performing.

High performance goes beyond output; it's shaped by how your team works, thinks, collaborates, and grows. As a tech lead, you play a critical role in enabling that growth.

UNDERSTAND THE FIVE DYNAMICS OF A HIGH-PERFORMING TEAM

Google conducted one of the most well-known studies on effective software engineering teams, known as Project Aristotle. Starting in 2012, Google spent two years studying 180 teams, 115 in engineering and 65 in sales, examining over 250 different team attributes. The goal was to identify what makes some teams more successful than others. Instead of focusing on individual talent, the study (*https://oreil.ly/mUG1N*) found that how team members interact with each other is the most critical factor.

Through this research, Google identified five key dynamics that define high-performing teams: psychological safety, dependability, structure and clarity, meaning, and impact. When I first came across these, I immediately reflected on how they applied to a team I led, one that was consistently described as high-performing.

Psychological safety

Having psychological safety in a team means that everyone feels safe to share ideas, challenge decisions, take ownership without fear of judgment or punishment, openly express their thoughts, admit mistakes, and ask questions without hesitation.

In our team, psychological safety meant we had a culture where speaking up was encouraged, and no one was afraid to challenge the status quo. Even though we came from different backgrounds and cultures, openness was a core part of how we worked. If something was bothering someone, they spoke up, whether it was about blockers, frustrations, or disagreements. They were vocal with me, with the team, and even with the client.

People asked "Why?," "Why like this?," "Why now?" constantly. Overcommunication became one of our most annoying and effective habits. We repeated things as many times as needed, reinforcing ideas and clarifying misunderstandings. There was no "You should already know this" or "Why are you asking this now?"

This openness extended beyond our team. Because we were so transparent, our stakeholders became more open and direct with us. Instead of vague conversations full of political correctness, discussions were clear and straightforward. We would hear things like "We need to go to market fast with this feature. We will pay the price of tech debt in the next iteration," and they actually followed through when we later said, "This time we need to do it right."

You can find more on this in the section "How to Create Psychological Safety on Your Team" on page 188.

Dependability

A high-performing team is reliable: things get done, up to the standard of quality, on time, consistently. This was one of the key pieces of feedback I kept getting from our stakeholders: "I can rely on you. I know things will either be delivered on time or I will know about any blockers or possible issues way ahead of time. I don't have to worry."

We were consistently delivering value, and this came from always challenging the value of the work we were doing and constantly asking the why behind it. Stakeholders knew that if we committed to something, it would either be delivered on time or they would be informed well in advance of any issues.

People would take initiative. When a blocker or a problem arose, they would reach out to other teams, talk to stakeholders, reach out to the whole company, making use of #general channels, just to get the thing done. Getting approval before acting was not our way. We were more of an "ask for forgiveness instead of permission" group. While this tendency tended to provoke my anxiety, as sometimes I felt like I was losing track of all the moving pieces, overall, their proactiveness and autonomy benefited me greatly; I never felt like a bottleneck in their way.

That reliability was fully embedded in how we operated internally. We could rely on each other as a group: we made sure no one was ever blocked for long. We used pair programming, knowledge-sharing sessions, and daily tech huddles to prevent knowledge silos. Standups weren't just status updates; they were about identifying blockers and helping each other move forward. We often said things like "It'd be good if another pair took a look at this; we've been stuck for three days and need a fresh perspective."

We extended that same mindset beyond the team. We kept dedicated Slack channels open with the teams who relied on us and made sure communication stayed consistent and smooth. We were always available to support others, and team members regularly jumped into conversations in other parts of the company when they had something valuable to add.

Even in a remote setup, we prioritized real-time conversations; for example, we preferred a quick call over a long Slack thread. Not everyone agrees with that approach, but I found that it created a sense of presence and dependability. People knew we were there, ready to talk things through. Of course, for this to work well, the calls had to be focused: we clarified the problem and outlined the key questions in advance to respect everyone's time. Whatever we discussed and clarified on the call was then shared back in the Slack thread so everyone stayed in the loop.

Structure and clarity

While adaptability was key, we also had well-defined processes to provide stability and clarity. Flexibility allowed us to respond to change, but structure ensured we weren't reinventing the wheel every time. Everything had a clear structure: delivery, deployment, onboarding, decision making, progress tracking, goals, ways of working, growth, and roles definition.

I am a huge fan of documentation, so I constantly encouraged my team to track changes on architecture diagrams, retro discussions, infrastructure changes (tracked through an infrastructure-as-code process), and decisions. This made onboarding a breeze, ensuring new team members quickly understood how we worked and their role in it. It also kept our day-to-day running smoothly; everyone knew exactly what was expected of them and what to focus on, which enabled efficiency and speed.

Meaning and impact

I grouped these two concepts together because they overlap: they both refer to the extent to which team members feel their work has a purpose and is making a difference, whether in the organization or society. We all want our work to be meaningful at some level, and we all know how important it is that people believe in what they do. When people feel connected to the purpose of their work, it has a massive impact on motivation and engagement. This is a game-changer for a tech team.

The mistake many companies make is involving developers too late in the process, only after the product has been defined. But in our case, our PM included us from step one, the initial brainstorming. We were part of the ideation phase, free to ask questions and challenge assumptions. This is crucial because developers are often assumed to be uninterested in the "why" behind a product, when in reality, they often don't see the value simply because they were never included in the conversation.

Being involved from the start has two key benefits. First, the whole team gains a deep understanding of why they are building something, who it helps, what problem it solves, and why it matters. This alignment ensures that every decision and line of code contributes meaningfully to the final product. Second, team members see a clear connection between their daily tasks and the bigger picture, making even routine work feel impactful.

Another common mistake is assuming that meaningful work applies only to customer-facing features, revenue-generating initiatives, or socially impactful projects. But different people find meaning in different things. In our case, we worked on maintaining and migrating a huge monolith. Even though it was the "money-maker" product of our client, it was still something that could feel tedious for developers. What motivated us wasn't its impact on revenue but the technical challenge of understanding this massive system, improving its quality, and migrating it without breaking anything. It was like solving a giant puzzle.

The same applies to platform teams. Because their users are internal teams rather than customers, many struggle to connect their work to the end product. But once they spend time with the teams using their tools and see the direct impact of their work, their motivation increases significantly.

Doing impactful work is just the first step. Equally important is making that work visible. I made it a habit to highlight my team's accomplishments and challenges in every possible setting. As a tech lead, you're often the only person from your team in the room, whether it's with clients, executives, or other stakeholders. That makes it your responsibility to represent your team effectively. Show up prepared, with a clear understanding of progress, open questions, and any blockers. Don't just report status; advocate for your team and make their contributions visible.

I encouraged my team to use every opportunity to make our work visible, starting with showcases. One of the most effective examples was our quarterly client showcase. While many teams treated it as a checkbox exercise, throwing together slides an hour before, I saw it as a powerful platform to highlight our impact, especially to stakeholders and teams we didn't interact with day to day.

So we treated showcases like any other deliverable. We added a dedicated preparation task to our board, prioritized it, estimated the time it would take to prepare, and planned it with clear expectations. We made sure to include context about our team, any changes since the last quarter, a rundown of accomplishments both big and small, the challenges we faced, and what was coming next. Everyone on the team was involved, and we ran dry runs to stay within the time limit and polish the delivery.

These sessions weren't just about recognition, though the positive feedback we received was motivating. They became moments of reflection and celebration, a chance for the team to step back and appreciate what we had achieved together. Over time, they also became a valuable way for team members to develop their presentation and storytelling skills, which helped build confidence and visibility across the organization.

This doesn't mean we were perfect; no team is. Every time you solve one problem, another appears. Overcommunication can be frustrating. Constantly challenging each other could get exhausting. And as a tech lead, you never feel like things are truly "smooth," because if you're focused on improvement, there's always something to work on. The key is keeping the bigger picture in mind, questioning why you're doing things the way you are, and being open to evolving your ways of working. What made us efficient was our honesty, our ability to talk things through, and our willingness to adapt, even when challenges came up.

In conclusion, a high-performing team is not defined by having the best developers, making no mistakes, or always meeting deadlines, and it's definitely not about being perfect. Instead, it's a group of people who care about each other, work well together, and have a healthy team culture, as the team results are deeply tied to how people collaborate, not just individual skills.

HOW TO BUILD A HEALTHY TEAM CULTURE

Team culture is how you work together, shaped by shared values, behaviors, and team norms. It's intangible and hard to measure, but it influences everything: how decisions get made, how fast you move, how people communicate, and how feedback flows.

While you can't track culture with a single metric, there are clear signs when something is off. In teams with a weak culture, trust is often missing. People hesitate to speak up, avoid risks, and hold back feedback. Collaboration drops.

Problems get pointed out, but no one steps in to solve them. Decision making slows, and progress grinds down.

In contrast, a strong culture feels energized. People feel safe, take ownership, offer feedback freely, and support each other. They challenge ideas without fear and focus on making things better.

Whether you like it or not, every team has a culture. The question is whether you're shaping it intentionally or letting it form by accident. If your team's dynamics don't reflect your values or expectations, the most effective way to change that is to shape the culture from day one.

One of the most powerful ways to do that is by aligning on how you'll work together. Every time a new person joins, the team changes. Don't assume everyone shares the same expectations. Instead, make them explicit.

A simple and powerful tool is a "Ways of Working" session. This is a structured conversation you run when forming a new team or when major changes happen. The goal is to agree on how the team will operate. This can include how feedback is given, how decisions are made, how often you run retrospectives, and anything else that shapes the team experience.

The intention of the session is surfacing assumptions. Write them down, talk through them, and listen for disagreement. You'll often uncover mismatches you didn't expect. That's where the value is. For example, one person might expect pull requests to be reviewed within a few hours, while another thinks 48 hours is totally reasonable. Or you might have different understandings of what "done" means; does it mean merged or fully deployed to production? Clarifying those early prevents future misalignment and gives you a shared foundation to return to.

Teams aren't static, so team processes shouldn't be either. Every time someone joins or leaves, you have a new team. Dynamics, needs, and expectations shift. What worked before might need rethinking. For example, when a new team member joined, we realized our 9 AM standup clashed with their child's school drop-off routine. We discussed it together and moved it to 10 AM. Small adjustments like this can go a long way in supporting inclusion, morale, and team performance.

This document becomes an anchor. When things drift, feedback sessions stop happening, or decisions start getting made in silos, you can return to what you agreed on and ask the team, "Do we want to change this? Or recommit to it?"

It also becomes a powerful onboarding tool. Instead of new hires guessing how the team works or making assumptions, they can see the team's expectations clearly from day one. It helps them integrate faster and more confidently.

You can even use this shared agreement in hiring. It gives candidates a clear sense of what it's like to work on the team and helps you assess fit more meaningfully.

As you define and evolve how your team works, it's critical to make sure those processes are actually working for everyone, not just the loudest voices or the most experienced members. Inclusion doesn't happen automatically. It has to be built into the way you run meetings, gather feedback, and make decisions. Every process you design, like retrospectives, standups, one-on-ones, should be intentionally inclusive. For example, when running retrospectives, use multiple ways to gather input: out loud, Post-its, anonymous forms. People have different communication styles, and a one-size-fits-all approach limits participation.

Here are some key processes every team should define and continuously revisit:

Onboarding
>   This helps new members integrate quickly and understand how the team works.

Hiring (if applicable)
>   The team's values and ways of working should be reflected in how you assess candidates.

One-on-ones between tech leads and each team member
>   These build trust, uncover blockers, and maintain alignment.

Standups
>   Daily or regular check-ins that highlight progress, blockers, and opportunities to help.

Progress tracking
>   A system that's visible and understandable to both the team and stakeholders. Avoid tools that are too technical or go unused. If nobody's using it, revisit or remove it.

Retrospectives
>   Structured reflection points to discuss what's working, what's not, and how to improve generating improving actions.

Even if your team is small or just getting started, having these foundations in place makes a big difference. They don't need to be complicated; they just need to be intentional. Every process you define contributes to shaping a team culture where people feel safe, supported, and set up to thrive.

HOW TO CREATE PSYCHOLOGICAL SAFETY ON YOUR TEAM

The first sign of a healthy team culture is that everyone feels safe to speak up, ask questions, and take risks without fear of judgment. This kind of safety is foundational to team performance. You can't even begin to address other team issues until it's in place. But, as important as it is, creating it isn't easy. In the following sections, I'll share some ways you can build it as a tech lead.

Act as an example

The best tool you have to shape your team's culture is yourself. If you expect people to behave a certain way, you have to consistently model that behavior. Your actions speak louder than any values slide or process document.

As a tech lead, you set the tone more than anyone else. How you give feedback, how open you are to criticism, how well you listen: these all send strong signals to your team. While psychological safety is a shared responsibility, your behavior sets the foundation.

Your team looks to you for cues, and the way you handle mistakes, feedback, and collaboration directly shapes how safe others feel to speak up and contribute. So constantly keep yourself in check: reflect on your behavior and actions and the impact it has on your team, and adapt if needed.

Normalize failure

One of the best ways to build psychological safety is to change how failure is perceived, both human failure (like making a mistake or not knowing something) and system failure (like production incidents or process breakdowns). Mistakes are inevitable in both areas. So instead of hiding them or punishing them, bring them into the open and make learning from failure part of how your team works.

Encourage your team to take ownership of their mistakes and admit when they don't know something. Support them in taking risks and using failure as a chance to learn and grow, instead of being afraid of how others might react. And the best way to teach this is by doing it yourself. Start by admitting when you don't know something. A lot of tech leads feel pressure to always have the answers, but showing your team that it's OK to say "I don't know" builds trust. If they see you do it, they'll feel safer doing the same. The next step is asking for clarification—"Can you please explain this again?"—instead of staying quiet. Chances are, someone else in the room is feeling just as lost.

Own your mistakes. Say, "I'm sorry, I made a mistake." Acknowledge what happened instead of pretending it didn't. Start small: share when you mixed up a date or used the wrong variable. As that becomes more natural, you'll get more comfortable being open about bigger issues too. This creates space for others to do the same.

Failure is one of the best opportunities for growth because we tend to stop and reflect only when things go wrong. For example, in one of my teams, a release unexpectedly took down a key part of the product. It was stressful at the moment, but the team came together afterward to run a postmortem. That session led to some of our best process improvements, from better automated tests to tighter release checklists, that we might not have prioritized otherwise.

Postmortems are a great tool for learning, but they work only if you run them with the right mindset. Instead of trying to find someone to blame, focus on what went wrong. What failed in the process? What were the conditions that led to the mistake? And how can we prevent it from happening again? The goal is to make sure it doesn't happen again, no matter who's involved.

Let's say someone broke production with a commit, and it took ages to revert because your deployment pipeline is slow. Instead of pointing fingers, use that moment to ask: how can we make this better for everyone? Maybe the pipeline needs to be faster. Maybe you need to add feature flags so you can turn things off quickly when something goes wrong. The point is to make recovery easier, reduce pressure on individuals, and improve the system, not to blame someone for triggering the problem.

Another great way to normalize failure is by creating a controlled environment where experimentation is encouraged and mistakes are easy to recover from. For example, having a strong set of automated tests, an integrated development environment, smooth rollback processes, and well-defined staging workflows that mirror production allows people to take action without the fear of irreversible damage.

Some teams even run "game days," structured exercises where a team member intentionally injects a fault into the system and the rest of the team conducts a mock incident response.

In one of my teams, we used a tool called Chaos Monkey, which was designed to randomly shut down services in our production environment. While it sounds risky, we used it carefully and intentionally to test the resilience of our systems and processes. It forced us to build more fault-tolerant architecture and improved our team's confidence in handling unexpected failures. You don't need to go this far to benefit from the same mindset. But aiming for this level of trust in your systems, where failure can be tested safely, is a goal worth working toward.

In a team where failure is normalized, people feel safer to experiment, speak up, and take initiative. That's how you build a team that learns fast, improves constantly, and supports each other through the ups and downs of product development.

Plan for failure, not just success

I remember this time my team was doing a huge migration from one system to another. Everything seemed ready, but I had a gut feeling something was off, something that stopped me from pulling the switch. I tried explaining my worry to the team, and we double-checked everything, but I still couldn't put my finger on what exactly was wrong. So instead of endlessly checking again, we came up with a different approach: preparing for failure as much as for success. We shifted our thinking from "How do we avoid something happening?" to a set of questions like "What if it goes wrong?," "Who will fix it?," "Can we roll back the change?" We made a plan: who would be on call if things went wrong overnight, who needed to be informed about the problem, what the financial impact could be. We even flagged it to our PM: "Something might go wrong, but we're ready to handle it."

Unfortunately, my hunch was right. Errors started rolling in during the night. It took us a while to fix them, but we were ready. We had a plan. And honestly, looking back, moments like that were great team-building exercises. Everyone was focused on solving a common problem, everyone was equally part of the decision, and it didn't matter who wrote the commit; it mattered that we agreed together to go for it.

It was also a great learning opportunity. We grew from it. We got better not just at shipping code but at thinking ahead, at supporting each other under pressure, and at owning the outcome together.

Also, by describing the worst-case scenario and having a clear plan to deal with it, you make people less afraid, because the biggest fear is often fear of the unknown.

How you deal with failure builds your character as a leader and shapes your team culture. So learning how to use it right benefits everyone.

Encourage honest feedback

Make feedback a key part of your team culture. It shouldn't be something you save for performance reviews; it should be part of how you work every day. When feedback becomes normal, people get used to sharing what's on their mind, what bothers them, and what could be better. It helps build a habit of honesty.

You can also encourage people to use feedback in their day-to-day interactions. When someone asks you how they can grow, don't just give them your input; point them to others who work closely with them and suggest they ask for feedback there too. Or, when someone comes to you with an issue they're having with a team member, instead of jumping in to solve it for them, ask if they've shared that feedback directly. More often than not, that conversation is exactly what's needed to move forward.

The more your team gets used to sharing feedback in real time, the more comfortable they'll be being honest with each other, and with you.

More on this in Chapter 5.

Encourage diverse perspectives

In a psychologically safe team, all perspectives are welcomed—not just tolerated but actively encouraged. As a tech lead, this means your team's day-to-day processes need to be designed with inclusion in mind. From onboarding to delivery rituals to performance reviews, every part of how your team operates should support a wide range of people, not just those who already "get how things work."

The best teams work because their differences strengthen them. When people bring different strengths, backgrounds, and perspectives, those differences can fill gaps and elevate the whole team. As a tech lead, your role is to create the kind of environment where that can happen: where differences are valued and used well.

This starts by asking the team how they work best. Get a sense of how each person prefers to receive feedback, what makes them feel comfortable sharing ideas, and how they like to make decisions together. These conversations might feel small, but they shape whether people feel heard, or invisible.

The onboarding experience is one of the earliest and clearest signals of whether a team truly supports diverse perspectives. Without a good onboarding process, the people most likely to thrive are those who've worked at similar companies before. Everyone else, especially those coming from different industries, backgrounds, or cultures, starts at a disadvantage. For example, without high-quality onboarding, engineers joining a fast-moving tech company from a non-digital-first company might struggle to find their footing in the first months and be more likely to leave.

Building a psychologically safe team also means creating space for all kinds of ideas, including the strange or unconventional ones. On my team, we used to hold brainstorm sessions where the only rule was that nothing was off-limits. Anyone could propose anything, no matter how wild it sounded. And that's exactly where some of our best ideas came from: the simple, elegant solutions that we wouldn't have considered in a more constrained or judgmental environment.

Of course, welcoming diverse perspectives also means learning to handle conflict. And while it might feel counterintuitive, conflict is actually a good sign. It means people care. It means they're comfortable enough to challenge each other and raise concerns. When everyone always agrees or, worse, no one speaks up at all, that's when you should worry. Healthy teams debate. They wrestle with decisions. And they emerge stronger because of it.

Inclusive collaboration also extends to technical practices like estimation. Too often, estimates are made solely by the most senior engineers on the team. But complexity isn't one-size-fits-all; what feels simple to a senior might be far more time-consuming for someone with less experience. If the task ends up in the hands of a junior engineer, will the original estimate still hold? Involving the whole team in estimation discussions helps surface these differences early, promotes learning, and results in more accurate, realistic plans. When you make space for open conversation around complexity, everyone's perspective has value.

Overcommunicate

Overcommunication was one of my high-performing team's most effective, and occasionally most annoying, habits. We repeated things constantly: goals, updates, questions, and answers. There was no "You should know this already" or "Why are you asking this now?" We'd rather explain something one more time than risk confusion or silent struggles. Everyone was vocal: to me, to each other, and to the client.

This openness brought many benefits but sometimes required some fine-tuning. I remember needing to have a few conversations with team members about adjusting how they communicated in front of clients. Not to silence them but to help them consider the delivery. New clients, especially, weren't always used to such direct communication. Still, the habit of overcommunicating, from asking "Why?," "Why like this?," "Why now?" to revisiting our goals or decisions, helped us avoid misunderstandings and supported clarity throughout our work. Even with these risks, I still believe the benefits of overcommunication outweigh the downsides, so I absolutely recommend it.

Build trust

None of the strategies mentioned before can be achieved without trust in the team.

Trust is built through constant communication, transparency, and follow-through.

Constant communication means having recurring checkpoints with your team through one-on-ones, or team alignment meetings like standups or plannings, and continuously sharing information and knowledge with them.

Transparency means being honest about blockers and possible issues, admitting to mistakes and when you don't know, and providing balanced feedback: both positive and improvement feedback.

Whatever you say or agree to doesn't mean much if you don't act on it. Actually, one of the simplest ways to build trust is this: say you'll do something, and then do it.

More strategies on building and maintaining trust with your team can be found in the section "How to Build Strong Relationships" on page 62.

Psychological safety must be inclusive. If even one person doesn't feel safe, the team as a whole isn't truly safe. It counts only when everyone feels free to speak up without fear of being judged, ignored, or penalized.

To understand how your team is really feeling, build regular check-ins into your team's rhythm. Retrospectives are a natural opportunity for this. Try starting with a quick safety check: ask each person to rate, on a scale from one to five, how comfortable they feel speaking up in the team. If several people respond with low numbers, that's a clear signal that psychological safety is lacking, which means the retrospective might surface the views of only the most vocal team members. In that case, pause and make psychological safety the main topic of discussion.

One-on-ones are another key tool. Use them not just to talk about work but to create space for people to share how they're really feeling. If you notice someone regularly holding back, withdrawing, or not contributing in meetings, bring it up gently and supportively. These conversations show that you care and that you're paying attention to more than just output; they help people feel seen and heard.

You can also use tools like the Psychological Safety Ladder Canvas (*https://oreil.ly/H6vPX*) that can help your team visualize how safe team members feel and where support is needed.

Encouraging collaboration also greatly contributes to psychological safety, as it builds trust, mutual respect, and shared ownership. More on this in the section "Enabling Collaboration Inside the Team" on page 209.

By showing vulnerability, encouraging open discussions, and encouraging a learning mindset, you create the foundation for a strong, positive culture.

WHAT HAPPENS WHEN PSYCHOLOGICAL SAFETY IS LOST

As hard as psychological safety is to build, it's incredibly easy to lose. You can spend months creating a safe environment, only to undo it in a moment. I've experienced this firsthand.

I was part of a team that felt genuinely close; we were collaborative, positive, and high-performing. We had been working together for a while, and we had a strong bond with our tech lead and good rapport with leadership. Everything felt solid.

Then one day, without warning, one of our team members was let go. No heads-up, no context, just that it was a "performance issue" and she had to leave immediately. The effect on the team was instant. Suddenly, it didn't feel like any of us were safe. People grew anxious, worried about their own performance, and afraid to speak up. Trust was shaken. The openness we had built disappeared almost overnight.

Whether or not the decision was justified isn't the point. How it was handled sent a message: you're replaceable, and no one will tell you why. Morale dropped. Engagement dropped. And it took us months to feel like a team again.

You can think of trust like social capital: every respectful interaction, supportive gesture, or fair decision adds to your balance. But it takes only one misstep, like ignoring team input or showing bias, to make a significant withdrawal. The more capital you build, the more resilient your team becomes. But the account is never infinite.

One of the quickest ways to damage psychological safety is through discrimination, even in subtle forms like microaggressions. Things like a casual joke or an offhand comment can land harder than you realize. They don't need to be overt to do harm. Culture is shaped by small, everyday actions, not grand gestures. So reflect on your own biases, stay aware of how your words affect others, and speak up when something feels off. If someone on your team is affected by bias or exclusion, check in with them. A quiet acknowledgment can mean more than you think.

Mistakes will happen, and will happen to you too. What matters most is how you respond. A simple, sincere apology goes much further than a defensive explanation. Avoid phrases like "I didn't mean to offend you" or "It was just a joke"; intent doesn't erase impact. Acknowledge the harm, take ownership, and show that you're committed to doing better. And when you witness something inappropriate, even a quiet "Hey, that didn't sit right with me" can help uphold the team's values.

Unconscious bias exists in all of us. As a tech lead, these biases can show up in how you assign work, who you go to for input, or whose ideas you elevate, often without you realizing it. For example, you might find yourself more frequently connecting with people who share similar backgrounds or communication styles. While this is human, it's important to recognize the pattern. Even subtle imbalances can lead to a perception of favoritism, which erodes trust.

To keep yourself in check, make a habit of gathering feedback from everyone on the team, especially from quieter members. Staying self-aware, questioning your patterns, and striving to be fair in every interaction are essential to maintaining a culture of psychological safety.

HOW TO EVALUATE IF YOUR TEAM IS HIGH-PERFORMING

High performance is more of an aim than a goal. It's about continuously pushing and striving to do better as a team. It's about continuously checking that you are on the right track. It's not something you can check completely off your list: you cannot get to a point where you say, "I have a high-performing team. It's done." Even when you get there, it takes continuous effort to keep the high standards.

You can't improve what you don't measure. But when it comes to evaluating things like psychological safety, collaboration, or trust, the metrics become more nuanced.

There are well-known frameworks and tools that attempt to measure engineering team performance, such as the following:

DORA (DevOps Research and Assessment)
> Focuses on deployment frequency, lead time for changes, change failure rate, and time to recovery. It's commonly used to assess software delivery performance.

SPACE framework

A more holistic model that includes dimensions like satisfaction and well-being, performance, activity, communication and collaboration, and efficiency and flow.

Accelerate framework

Closely tied to DORA, it provides research-backed insights into what drives high-performing engineering teams, with a focus on lean product development and DevOps practices.

While none are perfect, it's useful to be familiar with them, as they may be referenced in your organization.

These tools can be valuable, but they often require investment in data gathering, processes, and consistent interpretation. In the meantime, there are simpler, more qualitative signals you can use to evaluate how your team is doing and where you might improve:

Your own observations

Start with regular reflection. Amid the pace of delivery, it's easy to overlook signs of drift or tension. Block time to ask yourself, "What feels off in the team right now?," "What would I change if I could?," or "Is everyone contributing equally in meetings?" Look for quiet voices, disengagement, or recurring topics in one-on-ones or retrospectives that might point to deeper issues.

Team feedback

But don't rely on only your own lens; your interpretation could be biased or incomplete. Always bring these reflections into conversation with your team. Ask the team directly how they feel about working together. Retrospectives, one-on-ones, and anonymous surveys are all useful entry points. Go beyond surface-level questions. Instead of asking, "How do you think we're doing?" try "What's one thing we could improve about how we collaborate?" or "When do you feel most included or excluded in this team?" Use this feedback to validate or challenge your own assumptions.

Stakeholder feedback

External recognition is one of the clearest signs of a high-performing team. When stakeholders, whether PMs, execs, or other teams, trust you, they'll show it. That might look like positive, unsolicited feedback about your impact, stakeholders referencing your work to others, or being entrusted with high-visibility or critical projects.

If external praise is rare or vague, it could mean the team's value isn't well understood, or not as strong as it should be.

Reputation and interest

When your team is truly high-performing, others want to work with you. People will ask to join your team, shadow how you work, or borrow your practices. In my team, we saw this when engineers from other projects reached out for advice or expressed interest in rotating onto our team.

Engagement and ownership

Another good indicator is engagement. Are team members bringing energy and initiative to problems? Do they raise issues without being asked, follow through on commitments, and hold each other accountable? Or do you have to follow up repeatedly just to maintain momentum? A team's internal motivation is one of the most telling signs of its health.

Results and standards

Of course, performance also shows up in outcomes. Are you delivering what's expected? Are you hitting quality benchmarks? But results alone don't tell the whole story. It's the consistency, the collaboration, and the trust behind those results that really matter.

High performance is an ongoing pursuit. You don't "arrive" at having a high-performing team. Even once you've built one, it takes constant care to maintain. Priorities shift, people change, and new challenges emerge. Nobody out there has a team that's 100% high-performing, 100% of the time, and that's OK. Teams go through ups and downs, and part of your role as a tech lead is to keep checking the pulse, respond with empathy, and continually adjust.

How to Approach Performance Reviews

The goal of performance reviews is to have people reflect on their overall performance from the previous six months or one year and to give raises and promotions. It's also an opportunity to have people reflect on where they want to go and where they are and identify gaps in their skills and create a plan to address them. It's a chance to get feedback and give feedback, but it should not replace timely feedback (more on this in Chapter 5).

Based on your company structure and culture, your role in performance reviews for your team might be different. Some tech leads are barely involved in the performance review process for their team, as they are not expected to be involved. Other roles, like team leads or engineering managers, will take care of the heavy lifting of the process of review and growth.

And there are companies where you, as a tech lead, play a key role in the performance review process, given that the team lead role overlaps with your role and the engineering manager has a more high-level approach in the process.

My personal opinion is that the tech lead should play a key role in reviewing the performance of their team members, helping them build a case for promotion and acting as a supporter or cheerleader in front of management, or giving honest feedback when someone isn't meeting expectations. Who better to know about people's performance than you, who work daily side by side with them on technical and nontechnical topics?

Also, as a tech lead, I want to have a say in my team members' growth and development and do everything in my power to support them, as that will highly impact the overall performance of my team and, of course, me.

I have always been a key player in the performance review process for my team members, and this is the case that I will prepare you for in the next sections. I'll cover how to ensure everyone is aligned on the process and is fully making use of it, how to support them in the process, how to use different tools to measure performance, and what to do with the results.

But before we dive into supporting others, let's start with your own review process. Learning how to manage your own review well will help you better understand the system, set the right tone, and apply the same principles when guiding your team.

HOW TO PREPARE FOR YOUR OWN PERFORMANCE REVIEW

Your own performance review is the first process you need to manage. Leading by example will not only make it easier to guide your team through the process but also help you refine how you approach it. Instead of just telling them how to prepare, you'll be able to show them.

Get clarity on the process

Start by understanding how performance reviews work in your company. What are the timelines, expectations, and key steps? Who is involved, what tools are used, and how is progress tracked? Who makes the decisions, and who can support you? Some companies, like Thoughtworks, have the role of performance partners that act as mentors and advocate for your growth. Identify the people who can help you navigate this process, and make sure they are aware of your plans.

Gather all available documentation and list anything unclear. Then, go back to your manager to clarify assumptions and questions. Role expectations documents can also be useful, as they help you understand how performance is evaluated at different levels.

Once you have a clear understanding of the process, create a plan: define the actions you need to take, who you need feedback from, and how you will track everything. Many companies provide tools to collect and visualize feedback, but if yours doesn't, a simple spreadsheet will do.

Define what you want

What do you expect from your next performance review? Maybe you're aiming for a salary raise, a grade change, or opportunities to develop specific skills and connections. For example, I got most of my leadership training through a performance review cycle. I researched internal leadership programs and built a strong case for why I should be part of them, highlighting how they aligned with my development and the value I could bring back to the company.

Be specific about your goals. Whether it's getting a mentor, working on a particular skill, or connecting with peers in similar roles, having a clear ask increases your chances of getting what you want.

Build your lists

Start by listing what you're proud of, the impact you've had. As a tech lead, it's not just about your individual contributions anymore. Your success is measured by your team's success. Shift from "I delivered this feature" to "I helped Alan improve his soft skills through mentoring" or "I unblocked the team on an initiative by facilitating key conversations." Think about how you've enabled your team to do their best work.

> **Note**
>
> Consider keeping a running "brag document" (*https://oreil.ly/shufi*), as recommended by Julia Evans. It's a simple practice: maintain a private doc where you jot down accomplishments as they happen. That way, when review season rolls around, you're not starting from scratch, and you're less likely to forget the small, important wins.

Just as relevant as accomplishments are the challenges you've faced. The best learning often comes from things that went wrong. That's why postmortems exist, because reflecting on failures helps us improve. The same applies to performance reviews. Managers don't just want to hear that everything is great. They want to see growth and how you've handled setbacks, missed deadlines, skill gaps, and difficult situations. Take a look at what's happened since your last review: major milestones, wins, struggles, incidents, and feedback you've received. Use these to build a well-rounded view of your progress.

Make sure your accomplishments and challenges align with the expectations of your role. Many tech leads never revisit their job description; they simply assume they know what's expected. This mismatch often leads to frustration when aiming for promotions, as what they consider valuable may not align with what the organization or leadership is actually looking for.

For instance, some tech leads believe their main responsibility is only dealing with every technical challenge their team faces. As a result, they spend most of their time staying up-to-date on technologies and reviewing code while neglecting critical responsibilities like mentoring team members, enabling growth, or collaborating with product managers to shape a strategic roadmap. These overlooked areas are often exactly what the organization expects from someone in a leadership position. So, to avoid any bad surprises, make sure to validate your lists against the official role description and in conversation with your manager before moving forward. You can find more on this topic in the section "Understanding the Expectations of Your Role" on page 12.

Once you have your accomplishments and challenges, combine them into a cohesive narrative. Show how your wins and struggles contributed to your growth as a tech lead and the overall success of the team.

Ask for feedback

Feedback is a critical part of any performance review. You'll likely need input from your team, managers, and stakeholders. As a tech lead, this feedback is even more required because your performance is heavily influenced by how others perceive your leadership.

I've seen tech leads struggle because they felt they were doing poorly when their team actually thought otherwise or were caught off guard by unexpected feedback because they weren't checking in regularly. It's easy to misjudge your own impact when you don't have external perspectives. (For a detailed process on how to collect and use feedback effectively, check out Chapter 5.)

Use the feedback to refine your accomplishments and challenges lists.

Feedback is valuable only if you follow up on it. Have conversations to clarify points, gather additional examples, and resolve inconsistencies. Sometimes, different team members will have conflicting views of your role. Juniors may expect you to code more, while seniors appreciate that you focus on strategy and alignment. Understanding these perspectives will help you address concerns and set expectations.

Make your work visible

Use every opportunity in the review process to advocate for yourself and your team. Even if something is optional, consider doing it anyway.

At Thoughtworks, we had an optional presentation phase where people could present their case to senior leadership. Many skipped it because it wasn't required, but I always did it. It gave me a rare opportunity to tell my story, clearly state what I wanted, and address any concerns directly. Senior leadership taking the time to attend these sessions was a sign of great leadership in itself. These presentations didn't just impact my performance review; they helped build relationships with decision makers that benefited my career long term.

Performance reviews are a powerful opportunity to reflect on your impact, advocate for your growth, and shape your development path. As a tech lead, how you approach them influences your own progress and sets the tone for how your team engages with the process too.

HOW TO HELP YOUR TEAM PREPARE FOR THEIR REVIEWS

Now that you've seen how to manage your own performance review, it's time to apply that same mindset to supporting your team. Many of the steps, like reflecting on achievements, gathering feedback, and clarifying goals, overlap, but your role shifts. Instead of advocating for yourself, you're now helping others do the same.

As a tech lead, you play a key role in the performance review process, not just in guiding your team through it but also in evaluating their work. Your assessment carries weight, so it's essential to approach this process with intention and clarity.

If you're new to a team and have to conduct performance reviews without much historical context, acknowledge this up front. Let both your team and your manager know that your insights may be limited. In some cases, tech leads ask their engineering manager to handle the formal review while they focus on helping their team reflect on accomplishments and future goals. Even if you're not the final decision maker, you should still facilitate meaningful discussions to help your team navigate the process effectively.

Set the right mindset

First, make sure your team understands why performance reviews matter. Many techies see them as just another task to check off the list, missing opportunities to showcase their work, get valuable feedback, and advocate for growth. Without proper guidance, they often scramble to collect feedback last minute, leading to unnecessary stress and missed chances for meaningful recognition.

Take the time to communicate the purpose of the review process and how it impacts promotions, salary increases, and professional development. Clarify expectations, timelines, and what happens after the review. People are more likely to take the process seriously when they understand its value.

Make the process smoother

You can help reduce the friction of performance reviews by integrating them into your team's workflow well in advance. Plan ahead and factor in the time required for feedback collection, reflection, and discussions. This period will inevitably impact your team's delivery, so acknowledge that and adjust roadmaps accordingly. Keeping stakeholders informed of this impact helps manage expectations.

Make sure everyone knows the steps, roles, and responsibilities involved. Be clear about your role, what decisions you influence, what is outside your control, and how much weight your input carries. Ensure your team has access to all the necessary resources, templates, and tools, and encourage them to start preparing early. One-on-ones are a great way to bring up performance review prep gradually instead of dumping everything on them last minute.

Encourage your team members to think about their expectations for the process. Are they aiming for a promotion, a salary increase, or feedback on their growth? The earlier they articulate this, the better you can support them. If someone expects a 10% raise but the company's maximum is 5%, setting that expectation early prevents frustration. If someone is striving for a senior role, you can gather examples of when they demonstrated leadership or took initiative to strengthen their case.

Give effective feedback

Don't leave your feedback until the last minute. Some companies use automated tools for performance reviews, while others rely on individuals to gather input manually. Regardless of the system, aim to submit your feedback early and include specific examples. This not only strengthens your team's self-assessments but also gives them more insights into their own performance.

Before finalizing your evaluations, have one-on-one conversations with each team member. Go over your feedback together, highlight areas of agreement, and discuss any discrepancies. This reduces the risk of unpleasant surprises where people might read something later in your feedback they were not aware of.

Keep the process moving

Use your one-on-ones to check in on progress and ensure people are actively reflecting on their work, gathering feedback, and preparing their reviews. Running feedback-focused sessions, like Speedback (covered in the section "Building a Feedback Culture on Your Team" on page 132), can also help jump-start the process.

If you've been tracking team milestones and feedback throughout the year (as recommended in the previous section), this is the time to revisit those notes. Use them to remind people of their accomplishments and challenges, helping them build a more complete picture of their growth.

Help your team evaluate themselves

One of the most impactful ways you can support your team during performance reviews as a tech lead is by helping them evaluate themselves with honesty and clarity.

A great starting point is your company's official role descriptions or skills matrix. Use these as a shared reference to guide your conversations. Ask each team member to self-assess across the different expectations, either for their current role or the one they're aiming for. This will surface strengths, reveal growth areas, and help set a common foundation.

A 1–10 scale can be a useful tool here but only if used intentionally. Have them rate themselves across the relevant competencies, and then ask, "Why did you choose that number?" This simple question often unlocks insights they hadn't fully articulated. For team members who tend to undersell their contributions, walking through real examples can help them realize they're further along than they thought. For others, the conversation might highlight blind spots or the need to gather better evidence to support their self-assessment.

Then ask, "What would make this a 10?" This helps define what success looks like in practical terms. If you can't describe a 10, how will you know when you've arrived? I once had a junior engineer say their "10" in technical skills meant "being the best Scala developer on the team." But as she said it out loud, she realized that wasn't realistic, or necessary. She redefined it to something more grounded: "Being able to deliver a feature end to end in our Scala service on time and with minimal support."

Some people hesitate to ever rate themselves a 9 or 10, even when they meet the bar, because they fear appearing arrogant. I often say, "If you're never going to use a 10, why have it on the scale?" Or "Why should they promote you if you are not exceeding expectations in any area?" This might sound harsh at first but it usually triggers a shift in mindset, helping people assess themselves fairly.

Disclosure: this process isn't really about the numbers. It's about the clarity and reflection that come from justifying them and using that insight to shape what comes next.

Sometimes, you and a team member might not see eye to eye, especially when it comes to promotions. For example, many developers, especially at the senior level, fall into the trap of thinking that technical improvement is all that matters. But being a senior developer goes far beyond writing excellent code. It often includes mentoring others, leading initiatives, owning outcomes, collaborating effectively across departments, and demonstrating accountability. If a team member is focused only on developing their coding skills while working in a silo, they're missing key parts of their role, and that can hold them back.

These conversations can feel personal or even tense, but the solution is to move away from subjective opinions and focus on shared criteria. Refer back to the company's expectations and ask: What behaviors are visible today? What impact are they having? If they feel ready and you don't, walk through the expectations together. Point to concrete examples of where they meet them and where growth is still needed.

Handled well, this becomes a constructive, collaborative conversation. It's not "your opinion versus theirs." You're working together to evaluate what success looks like and how to get there. I've seen this go wrong when the only feedback someone receives is "I just don't feel you're there yet." That's when frustration and defensiveness take over. If you instead offer specific examples and show the path forward, most people will be open to, and even grateful for, the clarity.

Helping your team align their self-perception with the actual expectations of the role is one of the most empowering things you can do as a tech lead. It prepares them not just for performance reviews but for long-term, sustainable growth.

Answer last-minute questions

As the deadline approaches, expect a spike in last-minute questions and concerns. Keep your calendar slightly more flexible so you can provide guidance, set expectations, and help the team approach the process with clarity and confidence.

Helping your team through performance reviews means setting the tone, reducing anxiety, and creating opportunities for growth. By offering support, giving thoughtful feedback, and aligning self-assessments with company standards, you empower them to take ownership of their development and ensure no one is caught off guard when decisions are made.

WHAT TO DO AFTER THE REVIEW CONVERSATION

At this stage, your involvement becomes more limited, as the formal performance review process typically involves multiple managers. They take a high-level view, considering factors like budget, business priorities, and overall team performance. Their decisions about things like who gets promoted and who gets a raise are based on feedback, assessments, and company objectives.

As a tech lead, you likely won't have the final say on salary adjustments or promotions, but your input carries weight. Since you work closely with the team, your perspective on an individual's readiness for a senior role or their overall contribution will be valuable. Your written feedback is crucial, but expect to be asked by managers for promotion recommendations as well. Your insights can influence outcomes, even if the final decision rests with upper management. Any input you provide in these meetings should reinforce the guidance and expectations you've already communicated.

Once decisions are finalized, the results will be shared with the team. As always, some people will be happy with their outcomes, while others may feel disappointed. Be prepared for follow-up conversations; team members will want to understand what they need to improve and how they can grow.

The formal review process may end with final decisions and outcome conversations, but your role as a tech lead doesn't stop there. Whether you're supporting someone through a tough result or helping them plan their next growth step, your ongoing involvement is what truly drives development. Don't let the review be the last conversation. This is the perfect time to capture growth plans, while motivation is still high.

That means different things for different people:

- For those who've recently been promoted into a new role—say, from junior to mid-level or from senior to lead—help them settle into their new responsibilities. Review expectations together, identify stretch opportunities, and find ways for them to develop the new skills they'll need to succeed.

- For high performers or overachievers, don't just say "great job" and move on. Use their enthusiasm as fuel. Look for opportunities to stretch them, delegate impactful projects, and help them grow toward their next goal. This is a great way to keep them engaged and developing.

- For someone aiming for a promotion or a role change, work with them to map out a clear development path. Outline what success looks like, what gaps need to be closed, and how you'll support them along the way.

- For someone flagged as underperforming, you'll need to take a more hands-on approach. Collaborate on a targeted growth plan that addresses the key areas of improvement highlighted in their review. (You can find this process detailed in the section "Dealing with Underperformance on the Team" on page 216.)

No matter the outcome, the performance review is just a milestone in a continuous process. What happens afterward is just as impactful as the review itself. The conversations you continue, the support you offer, and the plans you help shape will define how each team member grows and how your team moves forward together.

MAKE PERFORMANCE REVIEWS A CONTINUOUS PROCESS

Performance reviews don't start with the official announcement to "get your review ready," and they certainly don't end with the final conversation about outcomes. A well-structured performance review happens all year long. Actually, the overwhelming nature of performance reviews comes from the fact that most people leave everything until the last minute: gathering feedback, compiling accomplishments, and preparing discussions.

But if you consistently track your accomplishments and challenges throughout the year, communicate them clearly to your manager, and use your one-on-ones effectively, to discuss goals, growth areas, and progress, then the actual review becomes a simple formality. By the time the review conversation happens, you should already have a clear idea of where you stand. And if something isn't possible, you should know well in advance rather than being caught off guard.

A simple trick I use, and one that everyone I've shared it with loves, is keeping a clear jar on my desk where I throw in colored Post-its with quick notes about things I'm proud of, milestones I've achieved, and challenges I've overcome. It's a low-effort way to track progress, and the visual reminder boosts confidence. Seeing the jar fill up over time serves as motivation, making it easier to recall accomplishments when review time comes.

If you prefer digital tools, a simple Google Doc works just as well. The key is to choose a format that's easy to update and review regularly, something you'll actually use.

The next step is to apply this same approach to your team, tracking their progress continuously throughout the year. Keep notes on key accomplishments, challenges, and feedback from stakeholders. Then, when review time comes, you won't need to scramble for details; you can just review your notes and put together a clear, well-supported assessment. This approach saves time, reduces stress, and ensures that nothing important gets overlooked.

This mindset should also extend to your team. If they regularly track their progress, performance reviews won't be a last-minute panic or a source of stress. There should be no surprises; people should already have a solid understanding of their strengths, areas for growth, and overall performance. A performance review should simply formalize what they already know.

Yet, as we all know, this rarely happens. Too often, people react with "Oh no, not again" when review time rolls around. As a tech lead, it's on you to make sure performance reviews aren't just a rushed, once-a-year event but an ongoing habit. They should be on everyone's radar every day, with tracking accomplishments, setting milestones, and having a growth plan becoming second nature. Every one-on-one is a great opportunity to talk about their career growth and skill development.

The good news is that most companies support ongoing performance management with structured tools for tracking feedback, growth plans, and individual progress. These tools make it easy to navigate feedback over time, as they capture who shared it, when, what was said, and any resulting action points.

As a tech lead, you'll often be the one ensuring that feedback is properly recorded and that it aligns with the expected format. It might feel like tedious admin work, but having structured, accessible records pays off, especially when it's time to write reviews, advocate for promotions, or support someone's development.

And as tech evolves, you may be able to offload some of that effort. For example, I know a tech lead who's experimenting with building an AI agent to input raw feedback into their internal tool, helping to save time while keeping everything organized. It's early days, but the idea of streamlining this kind of repetitive work is promising.

Constantly thinking about performance, tracking it, and measuring progress helps not only to simplify reviews but also to refine how you assess growth and impact. Like any evolving practice, performance reviews should adapt based on feedback and experience. Find ways to make them more effective for both you and your team. The goal is to turn reviews into meaningful, growth-focused conversations rather than a once-a-year stress fest. Done right, they become a powerful tool for both personal and team development.

Common Challenges and How to Overcome Them

Stepping into the tech lead role often means facing a whole new set of responsibilities, many of which no one really prepares you for. Suddenly, you're expected to lead hiring processes, onboard new team members, run performance reviews, and shape the team culture. These are big, high-impact activities that affect the team's success, yet most tech leads are left to figure them out on their own.

Throughout this book, I've covered many of these responsibilities, but some challenges consistently stand out, either because they're hard to get right or because they're easy to overlook until they cause real problems.

In this section, I'll walk you through three of the most common and critical challenges tech leads encounter when building and scaling teams:

- Building an onboarding process from scratch

- Enabling collaboration by building trust and alignment inside the team

- Dealing with underperformance, one of the toughest and most uncomfortable parts of leadership

If you've never handled these before, that's completely normal. What matters is how you respond. This section gives you clear, actionable strategies to tackle these challenges head-on, and help your team grow stronger because of it.

BUILDING AN ONBOARDING PROCESS FROM SCRATCH

If you're starting a team from scratch or don't have an onboarding process yet, here's a quick and easy way to put one in place with minimal effort the next time you hire someone new:

1. Assign them a buddy, someone who can answer questions and guide them through the first few weeks.

2. Ask them to document everything: the tools they use, the resources they find helpful, the challenges they face, and who they turn to for help.

3. Review their notes to ensure accuracy and completeness.

4. Help them publish it as an internal guide for future hires.

5. Give them credit! Recognizing their contribution makes them feel valued from day one.

6. Ask each new hire to use and update the guide as they go through onboarding.

That's it! You now have an onboarding process with minimal effort on your part, and the new hire gains early positive visibility within the team. This isn't the perfect onboarding process, but it's far better than having none. At the very least, your next new hire will have an easier time than the last.

There is no such thing as a perfect onboarding process, but there are definitely bad ones. As a tech lead, your job is to be intentional about improving and adapting your onboarding process. A well-structured, continuously evolving onboarding experience directly contributes to team performance, helping new hires integrate faster, feel more confident, and become productive members of the team.

ENABLING COLLABORATION INSIDE THE TEAM

We all know the benefits of proper collaboration, yet so many teams don't know how to work together effectively. Instead of complementing each other, sharing knowledge, and amplifying their collective impact, they end up getting in each other's way.

Some common symptoms:

- Knowledge silos; everyone depends on specific people for critical pieces of work

- Duplicated work; two people build the same thing, unaware of each other

- Time-wasting debates that go in circles and never lead to a clear decision
- Lack of ownership; work bounces around with no clear driver
- Reluctance to ask for help; people are afraid to look like they don't know
- Decisions made in isolation, then revisited later because others weren't included

If any of this sounds familiar, you're not alone. The good news is there's a lot you can do as a tech lead to encourage collaboration by designing team processes that rely on interaction and shared ownership.

Pair programming

Pair programming is one of the simplest ways to build collaboration into your team's day-to-day work. Mix up pairs regularly to break down silos, spread knowledge, and create natural mentoring moments, especially between juniors and seniors. But don't frame it as a process to enforce. Instead, introduce it as a solution. For example, suggest two people work together to debug a tricky issue or design a complex flow. That's often how it sticks.

That said, be aware that some people might resist it at first. On the surface, pair programming can feel like two people doing one person's job. It might seem inefficient, especially when deadlines are tight.

But the real value shows up over time. Pairing reduces the time lost in context switching or transferring tasks when someone's sick or goes on vacation. It lowers the overall work in progress, helping people stay focused on finishing instead of just starting. It naturally creates a "four eyes" check, so you often don't need a separate PR review. And maybe most importantly, it speeds up onboarding, improves knowledge sharing, and builds collective code ownership across the team.

Yes, there might be some short-term friction. But as a tech lead, you're thinking long term. Your goal is to help your team work more effectively together, and pairing is one of the most powerful habits to build that kind of culture, even if it takes some time to pay off and start seeing the benefits.

Knowledge-sharing sessions

There are a number of different ways to encourage knowledge sharing:

- Run mob programming sessions to demo new libraries or frameworks.
- Host informal tech talks: no slides required, just show and tell.

- Make demos a shared responsibility: instead of running them yourself, assign them on the team board and rotate ownership.

Support different styles; for example, some people prefer writing to speaking. Encourage those team members to write documentation, internal articles, or public blog posts (if there are no company restrictions).

Onboarding as a team sport

Onboarding is a great opportunity to get people to collaborate by simply using the buddy system and by making it a shared team process. This naturally gets people working together. See more ideas in the section "Involve the whole team" on page 176.

Distribute responsibility

If you notice that a few people in your team are owning multiple initiatives or holding onto all the critical knowledge, it's time to step in. This kind of concentration of responsibility often leads to knowledge silos, burnout, and slowdowns when those individuals are unavailable. It also limits growth opportunities for others on the team. Encourage people to delegate and share ownership.

Start by having open conversations with those who hold a lot of responsibility. Many times, they don't realize how much they're carrying or that it's actually blocking others from stepping up. Explain the broader benefits of delegation, both for the team and for themselves. For the team, it means learning, development, and stronger collaboration. For them, it means fewer context switches, less pressure, and possibly the freedom to take on something more exciting or strategic. You can frame it like this: "You've done a great job owning this area, and now that it's more stable, I think it's a great opportunity to bring someone else in to take over. That way, you can focus on that new initiative we discussed."

Make delegation feel like a growth opportunity, not a loss of control. Encourage them to mentor the person taking over. Ask them to document processes or walk others through the context. Make it clear that stepping back from ownership doesn't mean stepping away from impact; it just means their impact evolves.

Rotate facilitation roles. Don't let the same people always run retros, refine the backlog, or chase action items. Rotate those responsibilities to build shared ownership and reduce silent burnout.

Pair people on initiatives

One of the best ways to build confidence, accelerate growth, and encourage collaboration in your team is to pair people on initiatives, especially by giving less experienced engineers the lead, with a more senior person supporting them in the background as a "second chair." This model works because it gives junior or mid-level team members real ownership while ensuring they're not left to figure everything out on their own. They get the chance to navigate ambiguity, make decisions, and present ideas, all with a safety net nearby.

The key is to be crystal clear about the roles. Set expectations with the senior person up front: they're not there to take over or micromanage. Their role is to guide, mentor, and step in only when needed, allowing the less experienced lead to truly drive the work. Think of them like a copilot, alert and ready to support but not grabbing the wheel unless absolutely necessary.

For example, let's say someone newer to the team wants to lead a cross-team integration project. Instead of handing it off to your most experienced engineer, flip the script. Have the newer person take point, leading meetings, managing timelines, and coordinating stakeholders, while the senior team member shadows, provides context, helps with tricky decisions, and acts as a sounding board behind the scenes.

You should also make yourself available as a coach throughout the process. Encourage questions. Offer feedback. Help them navigate the process. Don't wait until something goes wrong; check in regularly to see how they're feeling and what support they need.

Solve problems together

Solving problems as a team makes great team-building exercises.

In one of my teams, we used to hold quick tech huddles after standup, gathering around a whiteboard to brainstorm solutions to tricky bugs or design challenges. No pressure, no formalities, just a space to throw around ideas and work things out together. These sessions helped us move faster, share knowledge, and stay aligned. The key is creating safety. Make it clear there are no bad ideas. Encourage everyone to speak up, even if they're unsure. Some of our best ideas started with "This might sound dumb, but what if..."

Postmortems are another great example. Run them with a mindset of learning, not blame. If someone pushes broken code to production, don't ask, "Who messed up?" Ask, "Why did this happen, and how can we make recovery easier next time?" Maybe your pipeline is slow or you're missing feature flags. Focus on fixing the system, not blaming people.

Premortems are also fun and useful. Before a risky launch, ask, "What could go wrong?" This surfaces hidden risks early and gets everyone thinking ahead.

If you have to plan a new technical solution that will require architecture changes and impact multiple systems, instead of leaving it to one person to figure out the plan, use a collaborative process like a TD (technical definition) or RFC (request for comments). This involves drafting a shared document that outlines the proposed approach and sharing it with all relevant stakeholders: team members, other teams, architects, staff engineers, and anyone else who might be affected. This format gives everyone the opportunity to ask questions, raise concerns, and offer improvements before any decisions are finalized.

Whether it's huddles, postmortems, or premortems, the point is simple: shift from "your problem" to "our problem." When teams solve things together, trust builds and collaboration becomes second nature.

Celebrate together

Dealing with problems as a team is critical, but so is celebrating wins together, and doing it consistently. It sounds obvious, but many teams skip this step. They move from one milestone to the next without ever pausing to acknowledge what they've achieved.

I used to be guilty of this too. Focused on delivery and always thinking about the next thing, I'd often forget to take time to celebrate. Fortunately, I had someone on the team who didn't forget. They'd remind us to pause, schedule a lunch, set up a ping-pong match, or just take a moment to appreciate how far we'd come, even for small wins. And it made a huge difference. That energy carried into the work, strengthened our team bond, and made us more resilient during tough stretches.

Celebration doesn't have to be big or time-consuming. Start small: end the week with a "what we're proud of" round during standup or in Slack, or create a #shoutouts or #gratitude channel to recognize individual contributions. Use retrospectives not just to talk about what went wrong but also to highlight what went right. And don't just celebrate results; celebrate helpful behaviors. Call out mentoring, unblocking others, stepping up when needed, or simply supporting a team member.

Normalize asking for help

One of the most effective leadership tools you have is vulnerability. Instead of expecting people to just do something, or demanding they do it, try asking for help.

This approach can be especially powerful when working with strong personalities: people who seem resistant to everything, who challenge every decision, and who always push back.

I once had a senior engineer who just would not get on board with a technical strategy the team had aligned on. I tried everything: walking through the pros and cons, looking for compromise, but nothing worked. He simply wouldn't budge. Meanwhile, I was under pressure from stakeholders to move forward. I didn't want to override him, so as a last resort, I tried something different. I got him into a one-on-one and said, "I'm going to be honest with you here. I really need you to get on board with this approach. I'm being pressured by stakeholders to make a decision today, and I'd really like to have you with us on this. I see your concerns, and the team is aware of them too. We're all willing to take on the risks together. So work with me here; what would it take for you to get on board?"

That conversation changed everything. He replied, "Thank you for sharing this with me. I didn't realize the pressure you were under or how much I was blocking progress. I'm OK moving forward with the team's decision. I'd just like to write down my concerns in case things go wrong so we have my ideas documented," which I readily agreed to.

That moment was a turning point for me as a leader. It showed me that asking for help, even from someone who disagrees with you, can shift the dynamic completely. It's not easy. It requires vulnerability, humility, and getting out of your comfort zone. But once I started doing it, everything changed. I saw my team jump in and take things off my plate without hesitation. When I admitted I was overwhelmed, or unsure, they didn't judge; they supported. And not just because they had to but because they wanted to. All I had to do was name the thing I was struggling with and ask.

Once I experienced that kind of support, I couldn't go back.

So if you're facing resistance or feeling stuck, try asking for help. Not as a tactic but as a genuine invitation. You might be surprised by how much it changes the conversation and the relationship.

Make information transparent and visible to everyone

If you want people to take ownership and collaborate effectively, they need access to the right information, without having to dig for it or ask around. Transparency builds trust, reduces misunderstandings, and empowers your team to take initiative.

Start by creating shared documentation spaces where everyone can track progress, processes, and decisions. Whether it's a project board, a shared roadmap, or a simple team wiki, make sure it's easy to find, easy to update, and actually used.

Before kicking off new projects, run short alignment check-ins. Use these to clarify expectations, constraints, roles, and responsibilities. It's a simple habit that prevents weeks of confusion later. When people know what's happening and why, they're more likely to contribute ideas, ask smart questions, and challenge things that don't make sense.

The more context you give your team, the more ownership they'll take.

You can find more information on this topic in the section "Be Transparent in Your Communication" on page 67.

Some of these approaches may take time to stick. Yes, pairing might feel inefficient at first. Delegating might slow things down short-term. But these are long-term investments. You're building a team that shares ownership, adapts faster, and grows together.

And remember, just because you create opportunities for collaboration doesn't mean people will use them right away. You need to lead by example. Use the practices yourself, talk about why they matter, and bring people along one step at a time. Not everything will work for your team, and that's OK. Try, adapt, and learn.

Hopefully, some of the techniques shared here will start to gain traction with your team, but in my experience, if those efforts aren't sticking, it's often because they're bumping up against bigger, underlying problems in the team's foundation. Two of the most common root causes are lack of psychological safety and unclear ways of working.

Lack of psychological safety is one of the biggest silent blockers to collaboration. When people are afraid of being blamed, judged, or seen as incompetent, they stop speaking up. They hold back ideas because they're afraid they'll sound stupid. They don't ask for help because they think it will make them look weak. They don't admit when they've made a mistake because they're afraid of the consequences.

So instead of open conversation, you get silence. Instead of learning from mistakes, you get cover-ups. And instead of collaboration, you get isolation. You can run all the retros, tech huddles, and pairing sessions in the world, but if people don't feel safe to be honest, those rituals won't work the way they're supposed to.

The second silent killer of collaboration is when the team doesn't have a shared understanding of how to work together. Everyone assumes they know what "good" looks like, but everyone has a different version. This leads to confusion around roles, ownership, and decision making. Tasks fall through the cracks, handoffs are messy, and accountability is vague. People step on each other's toes or, worse, avoid taking initiative because they're not sure it's their responsibility. You end up with the classic case of "everyone owns it, so no one does."

To dig deeper and address those root issues, I recommend checking out the section "How to Build a Healthy Team Culture" on page 185 and the section "How to Create Psychological Safety on Your Team" on page 188. That's where real, long-lasting change starts.

DEALING WITH UNDERPERFORMANCE ON THE TEAM

One of the hardest and most uncomfortable responsibilities of being a tech lead is dealing with underperformance. Because it's uncomfortable and hard to navigate, most tech leads just avoid it. They wait too long or miss the opportunity to prevent the situation altogether by not setting clear expectations early on.

Most of the time, when someone isn't meeting expectations, it's not because they're lazy or incapable, it's because expectations were never clearly set in the first place. We assume people know what we expect from them without actually saying it out loud.

So the first thing you should do is make sure every team member has a clear understanding of what's expected in their role. Use your one-on-ones to ask them to review the job description. Follow up on any questions or confusion. Just getting through that list once can eliminate "I didn't know" as a reason for underperformance. You don't want to end up in a conversation where someone says, "I didn't know you expected me to mentor others." That's a harder conversation than it needs to be.

The second thing: act fast when you see an issue. One common mistake tech leads make is waiting too long. You notice a problem; maybe someone is consistently missing deliverables or they're not as "strong" as you expected, but you hold off saying anything. And the longer you wait, the bigger the problem gets.

So as soon as you notice something, bring it up. It might be something they're not aware of, or maybe you haven't been super clear about what "good" looks like. If someone is supposed to be involved in interviews but keeps skipping them, ask them why. Is it because they don't know it's expected? Or maybe they're unsure how to do it and need help.

Sometimes it's smaller stuff, like communication. Maybe someone leading a team initiative isn't responding to questions from other teams on Slack on the subject. You assume it's obvious they should, but have you ever actually said it out loud? Have you told them that responsiveness is part of leading an initiative? A quick conversation can clear that up.

Whatever it is, give feedback fast. The value of feedback drops the longer you wait, especially when it's about something that needs improvement. If you feel like someone is struggling across multiple areas, go back to the expectations and talk through where the gaps are.

I've seen a lot of tech leads put off these tough conversations, thinking they'll wait until the next performance review. But that's usually months away, and by then, it's too late. Telling someone "You could have handled that client call better...six months ago" just doesn't help. It's not actionable anymore; it just feels like blame. And that kind of delay can mess with someone's career path. A simple, timely bit of feedback early on could've helped them course-correct before things got worse.

Let's say you've been quietly hoping someone improves their code quality. You hint at it in retros with comments like "We need to get better at testing," but nothing changes. Fast forward a few months, and now you're having a hard conversation: "Your work keeps coming back from QA with bugs. We need to talk about a performance improvement plan." At that point, they're blindsided, and honestly, they're not wrong to be. A simple, specific conversation early on could've made a huge difference. I've seen people completely turn things around after one honest and well-timed piece of feedback.

Supporting someone's growth, especially when performance is falling short, can be a complex and emotionally demanding process. That's why you shouldn't do it alone. If you're seeing persistent issues or if it's unclear how to proceed, involve your engineering manager, HR, or the people team early. Managing underperformance often comes with legal and procedural implications, things that are completely out of your control and outside your responsibilities.

Before taking any action, make sure all the people involved agree there's a challenge worth addressing. This includes you, the person in question, and the rest of the people involved mentioned before.

Once there's alignment that something needs attention, you can jump in to help from your role. This isn't about creating a formal performance improvement plan, which is typically owned by HR or the EM, but rather cocreating a practical, supportive plan to help someone improve. You are the closest to this team member in the day-to-day, so you are in the best position to help them shape a growth plan and take action.

Start by setting the foundation. Help the person understand that growth isn't about being perfect; it's about making continuous progress. Encourage a growth mindset, frame challenges as learning opportunities, and remind them that improvement is possible with effort.

Help them self-assess. For this, go back to the expectations of their role and evaluate—What are they already doing well? Where are the gaps?—going point by point. Once their assessment is done, provide your feedback on where they are and also ask them to collect feedback from the rest of the team members to ensure their assessment is accurate.

Then define a shared goal: "How will we know we've made progress?" Break it down into smaller, actionable steps. These could be tasks, habits, or skills they need to build. For example, if the goal is to improve communication, a milestone might be taking over the communication on a new initiative that involves working with another team.

Set a loose timeline. Don't overengineer it, just enough to create momentum and accountability. Milestones might span a few weeks or a couple of months, depending on the nature of the issue. From there, talk about who can support them. This might be you, other senior team members, a mentor, or even external coaching. Let those people know so they can offer guidance and help reinforce the plan.

Just remember: support doesn't mean taking over. At the end of the day, it's still their responsibility to own the plan and follow through, so make that clear. You can create space, give guidance, and offer feedback, but you can't do the work for them. Improvement takes personal commitment. That's something no one else can provide on their behalf.

Also, look for stretch opportunities: are there projects or responsibilities that would help this person practice the skill they need to grow? Are there books, courses, or other resources they can use? Make sure they have access to the tools they need to succeed.

Finally, have a common way to track their progress. This will help with reflection and adjustment. Use your one-on-ones to check in on progress. Celebrate wins, talk about blockers, and be willing to adapt the plan as things evolve. The key is to make it a living document, not something you agree on once and forget about.

This tracking method also serves another purpose: it becomes your record of the steps taken. Documenting what's been tried—feedback shared, support offered, goals set, and check-ins held—ensures there's a complete picture of the situation. This helps the business make informed decisions if things escalate and reduces the risk of legal or procedural issues later on.

You can find a much more detailed breakdown of each of these steps in the section "Developing a Personal Growth Plan" on page 38. That section is written for tech leads building their own growth path, but the same ideas apply perfectly when helping someone else grow too.

As you support their growth, keep your manager or HR partner in the loop. In some cases, the EM may want to co-own the plan or track progress directly. They'll also help decide what happens if things don't improve. That next step might involve moving toward a more formal process, such as a performance improvement plan (PIP), to ensure fairness, alignment, and compliance with company policies. The PIP provides one last structured opportunity for improvement with clearly defined goals, timelines, and outcomes.

Eventually, it may become clear that it's time to pull the plug and let someone go. This is one of the hardest processes you might face as a tech lead, but sometimes it's the only path left. Keeping someone in a role they're not suited for, despite support and second chances, not only continues to affect the rest of the team, but it can also hold the individual back from finding an environment where they could thrive. Letting someone go should always be done with compassion, transparency, and support, but avoiding it when it's clearly needed only prolongs the harm to everyone involved.

Dealing with underperformance is never easy, but it's part of the job. When handled with care, honesty, and the right support, it can lead to real progress and prevent long-term damage to the team.

Key Takeaway

Building and scaling a tech team is one of the most rewarding, and challenging, parts of being a tech lead. From recruiting and onboarding to running performance reviews and building team culture, you're responsible for shaping how your team grows, collaborates, and succeeds together.

The work isn't always straightforward. You'll face ambiguity, tough conversations, and growing pains. But with the right mindset, clear systems, and a commitment to continuous improvement, you can create an environment where your team thrives.

Whether you're preparing your first hiring plan or dealing with underperformance for the first time, I hope this chapter gave you the tools to take action with confidence, and reminded you that you don't have to get it perfect; just be intentional.

8

Addressing Technical Challenges

Throughout this book, I've focused heavily on the people side of tech leadership, and for good reason. Trust, alignment, and collaboration are what make teams effective.

But as a tech lead, your influence also shows up in the technical foundations your team works on every day. Architecture, deployment, testing, incident response, technical debt: these are leadership concerns too. The way you shape and evolve your systems has a direct impact on delivery speed, product quality, and team morale.

This chapter is about those critical technical challenges.

I'll start with architecture: how to lead your team through design and improvement work, whether you're building from scratch or evolving an existing system. Then I'll dive into the mechanics of modern software delivery, including integration, deployment, and testing, and how to make them smoother and more reliable. Finally, I'll tackle some common but often overlooked challenges: managing tech debt, responding to incidents, and deciding how and when to document.

If the rest of the book has helped you build a strong team, this chapter will help you make sure that team is set up to build strong systems.

Architectural Strategies and Implementation

Architectural decisions shape everything a team builds, scales, and maintains. As a tech lead, you need to actively guide how your architecture looks today and how it grows and evolves over time.

In this section, I'll cover both ends of the spectrum, from defining a brand-new architecture from scratch to improving and evolving an existing one.

You'll also learn how to make architecture more visible and actionable through clear diagrams, how to define and manage cross-functional requirements, and how to navigate the trade-offs between innovation and stability.

DEFINING A SYSTEM ARCHITECTURE FROM SCRATCH

While most engineers work on existing systems, there are moments, like those involving greenfield projects, new products, or internal tools, where you get to define the architecture from the ground up. These are rare opportunities to shape a system intentionally from day one and set it on the right path.

Before jumping into code, take a step back with your team to answer foundational questions. Start with the problem: what are we solving, and why does it matter? These conversations should include product and business partners. Establishing clarity early ensures that you're designing the right thing. It also helps prevent overengineering, since you'll focus on solving a clearly defined problem.

Once the problem is clear, examine your constraints, technical, organizational, and business-related. These include decisions like which programming language to use, which cloud provider to adopt, or whether you'll follow certain internal platform standards. Make trade-offs visible. If you're optimizing for speed now, be honest about what you're deferring and plan to revisit later.

With this context in place, begin shaping a high-level technical vision: your early hypothesis of what the system might look like as it grows. This doesn't have to be precise. A rough sketch of the major components and how they interact is enough to align the team and prompt discussion. Questions like the following will help guide you:

- What are the core components or services we'll need?
- How should those components communicate or integrate?
- Are we anticipating shared infrastructure or reusable capabilities?

Then shift from the abstract to the actionable. Focus on the first working version: what must exist to deliver value quickly? Zoom into that initial slice of architecture:

- What is the first component or feature to build?
- What dependencies will it have?
- What tools or technologies will we use (e.g., databases, messaging systems)?

You don't need detailed diagrams. Even a simple C1-level diagram from the C4 model (more on the C4 model approach in the section "Visualizing a System Architecture" on page 226) can create a shared mental model that avoids misalignment and costly rewrites later. It also makes it easier to communicate your thinking across engineering, product, and other stakeholders.

As the system evolves, continue to move between the vision and the current state. Revisit your assumptions, update your diagrams, and document the decisions you're making. The point is to be deliberate about shifts in direction, and transparent about their implications.

It's OK if things change. They will. What matters is that the team stays aligned on what's changing, why it's changing, and what it means for the product and business. That's what a strong architectural foundation enables.

IMPROVING AN EXISTING SYSTEM ARCHITECTURE

Most engineers are working within systems that already exist: codebases with history, architecture shaped by past decisions, and tech debt that comes with real-world complexity. These systems need constant improvement, and that means their architecture must evolve over time.

As a tech lead, your role is to ensure changes happen for the right reasons. Resist the temptation to chase the latest trends or rewrite systems just for the sake of novelty. Start by asking questions: What real problem are we trying to solve? Are we facing performance issues? Struggling with integration? Are delivery bottlenecks slowing us down? Aligning on the true problem will help your team make better decisions, and stop you from changing everything unnecessarily.

Also, keep in mind that every architecture choice comes with trade-offs. No design is perfect. Part of your role is to make those trade-offs visible, to your team and your stakeholders. Improving architecture always costs something, whether that's engineering time, focus, or added complexity. Be clear about what you expect to gain, make sure the team has the space to do the work properly, and be ready to measure whether the change actually delivers the promised value, whether that's better delivery speed, improved stability, more scalability, or reduced operational burden.

For example, there are developers who are very keen on microservices being the best and only way to go for architecture. But who is to say that a monolith is fundamentally bad? The value of microservices comes when you need distributed ownership, when your product is scaling, or when you need independent deployments to speed things up. But this doesn't mean a monolith is wrong. Actually, every startup idea usually begins as a monolith: a simple service solving a specific problem. Microservices become necessary only when scaling demands it.

Passionate, deep technical developers often get caught up in big changes: massive refactorings or endless architecture updates. I prefer to stay agnostic about specific approaches and focus on what fits the problem best. Spend more time understanding the problem and the current constraints than chasing shiny new ideas. Once you jump straight into solution mode, it's easy to lose sight of the real goal.

Before jumping into any architectural change, pause and ask: What happens if we don't change this? or Why is this change necessary now? These questions help ground the conversation in real needs and avoid you falling into the "trend-driven development" trap.

Once there's alignment that a change is needed, don't assume it has to be all or nothing. Ask: do we really need to change everything at once? Often, a smaller incremental change is easier to execute, and easier to sell to stakeholders, than a risky big-bang rewrite.

Here's an example from one of my teams where the big move seemed tempting but we chose a different path.

The problem

We had a product living inside a large monolith deployed in a datacenter. We wanted to run some quick experiments on it, but that wasn't easy. There were multiple challenges that made this difficult.

First, the business logic was tightly entangled with other features in the monolith, making any change risky. We couldn't confidently update one part without potentially breaking something else.

Second, the tech stack itself added friction: .NET with C# code, RabbitMQ, a massive MySQL database, all hosted in a datacenter. Deployments were slow and painful.

At the same time, we had a newer environment that the team used for more recent products, built on a Scala microservices architecture, running in the cloud with Kafka streams and DynamoDB. It was designed for flexibility and fast iteration.

The solution

It was tempting to start rewriting everything in the new stack. But there were major problems with that approach.

First, it was a huge effort, impossible to estimate reliably. We could have thrown a "six months" estimate at it, but that would just have been a number with no real basis. Plus, during that time, we wouldn't be able to add any new features, which was, of course, unacceptable.

Second, blindly duplicating everything meant also migrating unused features and tech debt, something that we were really trying to avoid.

So we asked: What's the real reason for the move? Why now? It wasn't an easy discussion, but eventually we pinned it down: we needed to improve the main ranking system, the automatic classification ranking, which was crucial for the business. In the monolith, it was very hard to experiment with changes to it. That was our real driver.

Instead of migrating everything, we started by isolating and moving just the ranking feature. This approach is a classic example of the strangler fig pattern, a gradual migration strategy where new components are built alongside the old system, eventually replacing it piece by piece.

We deep-dived into how it worked. We added tests at every level. We isolated the feature from tangled logic. Unsurprisingly, we found dead code and several gaps between the actual behavior and what the product team thought it was doing.

Once we understood exactly what needed to work, we started building the new version: a fresh Scala microservice with its own DynamoDB database, connected through Kafka streams. After setting up the service, the ranking logic itself wasn't hard to reimplement. The real challenge was verifying it behaved identically to the old system.

For a period, both systems ran side by side. Every night, the same data would be processed by both the old and the new system, and we compared the results to ensure perfect alignment. Once we were confident the new system matched, we cut over. All ranking requests went through the new environment, although we still updated the MySQL database in parallel, keeping the old system's dependencies alive.

Yes, there were downsides: increased complexity, higher operational load, duplicated infrastructure costs. But it met our main goal: we could now experiment fast with the ranking logic and had already started applying machine learning improvements. On top of that, we now could deploy changes safely and ship in small increments.

The takeaway

Meaningful architectural change doesn't have to be massive. Start small. Focus on solving real, current problems. Anchor decisions in clear goals, and always understand the "why" before diving into the "how."

Tip

If you want to dive deeper into how to design systems that support change over time, *Building Evolutionary Architectures*, 2nd edition, by Neal Ford, Rebecca Parsons, Patrick Kua, and Pramod Sadalage (O'Reilly) is a highly recommended resource.

VISUALIZING A SYSTEM ARCHITECTURE

Before you can improve your system, you need to see it.

Yet many teams skip this step.

I've definitely been guilty of postponing architecture visualization for too long. It's easy to get caught up in building features, fighting fires, or tackling whatever's directly in front of you. High-level work like this often slips through the cracks, especially when it's hard to explain its value to stakeholders. Until you've done the work, it's difficult to show how it will positively (or negatively) impact delivery.

But if you want to make sound technical decisions, plan improvements, or add more load to your system, you need a clear view of the current state. You need to understand all the moving parts.

My mistake was making the task too big. I'd assume I needed to review all existing diagrams, verify every system's current state, and create everything that was missing, all at once.

The key is to start small. Break the work into manageable steps. Build a plan with your team to gradually develop visibility, and treat it as part of your ongoing delivery work.

First, check if there are any existing versions of architecture diagrams. Sometimes you'll find bits and pieces scattered across different places that you can build on. Other times, there might already be a full diagram of the system that just needs updating. If you find one, approach it with empathy. It's easy to look at an existing system and immediately start criticizing: "How could people make these decisions?" but always keep in mind the prime directive: "The team did the best they could with the resources and time they had." Even if it's not the best, at least you have something to start from.

If there is no diagram of your current system, start building one. The simplest way to start is by using the C4 model approach, developed by Simon Brown, that helps you visualize systems at different levels (Figure 8-1):

C1: Context
    The big picture, a high-level overview

C2: Containers
    Applications and data stores

C3: Components
    Internal structure of containers

C4: Code
    Class diagrams

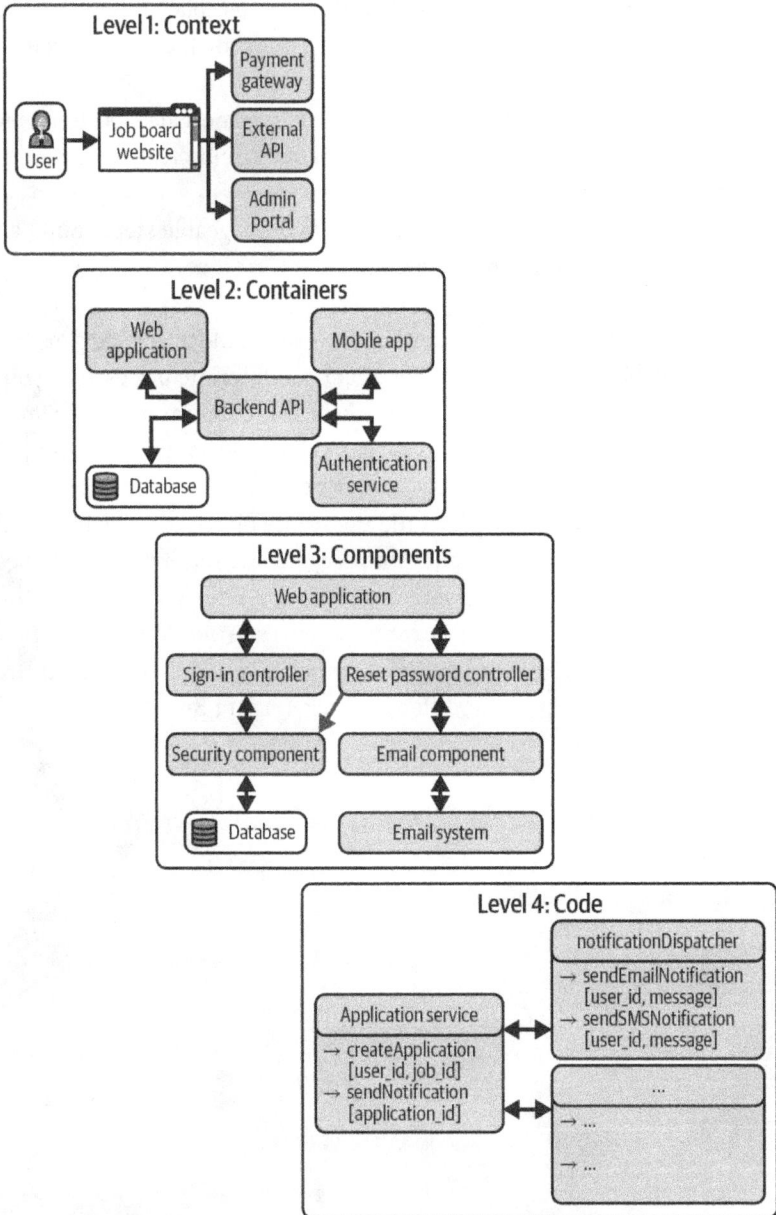

Figure 8-1. C4 model example

Each level of the C4 model describes a different level of detail of a system architecture:

C1: Context

The C1 level shows the system in its environment: how your product interacts with users, external systems, and other teams. Start as simply as possible: boxes for the pieces you know, one service, one database, one entry point, and build from there. Ask your team to contribute and keep adding pieces. Every time you add a new feature or discover a new dependency, update the diagram.

C2: Containers

The C2 level shows the main applications and databases in your system and how they interact.

Use it when your system involves multiple services or data stores. It's especially helpful for architecture reviews, planning deployments, or explaining runtime interactions and tech-stack choices.

C3: Components

The C3 level shows the internal structure of a single application—its main classes, modules, or functions—and how they collaborate.

Use it when refactoring, reviewing technical debt, or onboarding developers into a specific service. It's also useful when ownership is unclear or responsibilities are tangled.

C4: Code

The C4 level shows class-level design and relationships between individual classes or functions.

Use it rarely, only when working in complex legacy codebases or regulated environments or when a detailed view is critical for debugging or audits. Most teams do just fine without this level. It's costly to maintain, and if your code is clean, readable, and well-structured, the added value is usually minimal.

Tip

For a deep dive into this C4 model topic, check out the C4 model website (*https://c4model.com*).

Start with C1, then go deeper only as needed. A good rule of thumb is to dig further when the current level of detail no longer helps you make decisions, clarify confusion, or explain the system to others. Sometimes, the C1 level might be enough to drive technical decisions internally and explain to your stakeholders the impact of their requests and why timelines may need adjusting.

I once worked for a client migrating all their services from a datacenter to the cloud (Amazon Web Services/AWS). It was a huge company, lots of teams, and many interdependencies. My team owned a big monolith, the money maker, tightly coupled to a huge common MySQL database and many other services from other teams. We had to find a way to move this, piece by piece.

This required all teams to have an overview of their services and, most importantly, their interdependencies. I remember we built this huge diagram on a Confluence page that, when zoomed out, was unreadable: lots of services, databases, Kafka streams. Someone even tried drawing it on a wall-sized board, and it still didn't fit.

But every time we discussed a migration or someone asked how the monolith depended on other services, we went back to the diagram. Initially, it showed only what we knew. But with every conversation with other teams, we updated it, broadening our perspective.

While the C4 model is a great framework for visualizing architecture across different levels, it's not the only way to break things down. Another effective approach is to organize your diagrams around slices of functionality. For example, instead of one massive, unified view, you could create separate diagrams for things like checkout flow, authentication, or inventory management. Each of these can still share common components, but isolating them helps clarify purpose and reduce cognitive load.

Whether you use C4, functional slices, or your own framework, what matters most is making your system visible. Once you can see it, you can evolve it.

As teams increasingly adopt AI coding assistants, these tools are opening up new ways to generate helpful visualizations on the fly, without heavy up-front investment.

For example, you can use AI tools like coding assistants or chat-based interfaces to quickly generate sequence diagrams using diagram-as-code formats like Mermaid. These can be incredibly helpful when exploring an unfamiliar codebase or debugging a tricky call stack. Manually creating sequence diagrams is usually a waste of time; they're tedious and tend to go out of date quickly. But when generated quickly for one-time use, they become a powerful tool to build understanding in the moment. You don't even have to store these; just treat them as throwaway tools to help you grasp what's happening in the code more easily and move forward faster.

One particular scenario I have in mind here is a project I worked on that was built around a complex state machine. Explaining how the system worked to new joiners was always a bit of a struggle; even I had a hard time when I first joined. If I'd had an AI tool back then that could analyze the code and generate a visual diagram, showing the different states and transitions in a clear, structured way, I would've absolutely used it. It could have helped me trace how state changes flowed through the system and made those onboarding conversations far more effective.

Even if these diagrams aren't 100% accurate, they can still be extremely valuable. They give you a rough map to guide exploration or uncover misunderstandings. A "good enough" diagram can often move you forward faster than no diagram at all.

Of course, there are limitations. If the codebase is large or you try to visualize the full system, the output can quickly become overwhelming or inaccurate. These tools are most effective when you focus the scope, just enough to answer the question at hand.

Just like sequence diagrams, C4 diagrams can also be generated with the help of AI tools. You can ask ChatGPT or GitHub Copilot to generate a simplified C4-style representation of your architecture directly from code. For instance, you might paste in a few key files and ask, "Analyze this codebase and generate a C4 container diagram in Mermaid format. Assume it's a microservice that processes orders, with a database, an external payment provider, and a frontend client."

These tools can offer a useful starting point, especially when working with legacy systems or unfamiliar codebases. But they're not perfect. Code rarely captures every architectural decision. Responsibilities, boundaries, and implicit contracts are often missing or ambiguous. As a result, AI-generated diagrams may be overly detailed, skip important relationships, or lack the broader context needed to interpret them effectively.

The best results come when you combine AI-generated drafts with team input. Use the AI output to trigger conversation, then refine it collaboratively. Store the final version as diagram-as-code (using tools like Structurizr DSL or C4-PlantUML) so it stays versioned, reviewable, and easy to evolve.

In the end, the point of all this isn't to produce diagrams for their own sake. These tools, whether manual or automated, are valuable only if they help you reason more clearly, communicate more effectively, and make better decisions with your team.

DEFINING AND MANAGING CROSS-FUNCTIONAL REQUIREMENTS

Cross-functional requirements (CFRs) are architectural characteristics, and they are at the heart of every system we build. They shape the structure, the behavior, and the experience of using our software, whether we notice them or not. When we talk about architectural characteristics, we're talking about things like structure (the parts and how they fit together), what makes the system unique, what its purpose is, and what decisions along the way shaped it.

The most common CFRs you'll run into are the famous "-ilities": accessibility, compatibility, scalability, observability, reliability, and availability, but also performance and cost.

There are a few things these characteristics all have in common: they're nonfunctional (not directly visible to users), they're cross-cutting (impact the whole system), and they're much harder to test compared to normal functional features.

You might have heard CFRs referred to as *nonfunctional requirements*, but I prefer *cross-functional*, as *nonfunctional* makes them sound like an afterthought, and that's exactly how they often get treated, until something blows up.

Sometimes, the business itself makes these CFRs crystal clear. If you're working for a bank, security is nonnegotiable. It's baked into every decision, even if it slows development down. Or think of a high-frequency trading platform, where performance and speed are literally the business.

But most tech teams aren't that lucky. Most of the time, CFRs are invisible, until something breaks. Until your website loads too slowly and customers leave. Until your API falls over at 100 requests per second. Until you get hacked because security was "a later thing."

You already have CFRs, whether you think about them or not.

Every system has an expectation for speed, reliability, and resilience. The danger is assuming everyone's "common sense" matches. If you don't talk about it, you leave it open to interpretation. One person's "fast" is another's "barely acceptable." One person's "scalable" is another's "laughably small." That's how you get into trouble. Product says, "This feature feels slow." You say, "It's not." And both of you are technically right, but you weren't actually aligned.

The earlier you define CFRs, the fewer fights and surprises you'll have down the road. Ignoring CFRs can lead to expensive and painful retrofits. Things like system performance, resilience, and observability are way harder and more expensive to deal with later. Better to make them visible and agree on them early.

This is exactly where SLOs (service-level objectives) and SLIs (service-level indicators) come into play. They're just ways to make CFRs explicit and measurable.

So start defining them. You don't have to drop everything tomorrow and make a huge CFR playbook. Start small. Capture what you already have: current API response times, typical request volumes, average page loads, known security gaps. Write it down. Share it with your team. Share it with your PM. Validate assumptions. You might surface important misalignments right away.

Bring CFRs into everyday conversations. When planning a new feature, ask how many users it needs to support. Add load testing to your major releases. When integrating with another team's system, define what "good enough" means on both sides.

Document whatever you find somewhere visible and versioned. Observability dashboards work great for this. If you have monitoring tools like Datadog or Grafana, use them to track CFRs like latency, throughput, error rates, and uptime. Where it matters most, define real SLOs and alerts. For example:

- SLO: 99.9% of API requests should complete in under 300 ms over a 30-day window.

- Alert: Page on-call if 99th percentile latency exceeds 500 ms for more than five minutes.

SLOs are powerful negotiation tools. They let you say, "We can't take on this new feature without risking our performance guarantees."

Talking about CFRs early also helps with making real trade-offs. You can't have perfect scalability, reliability, performance, security, and cost optimization all at once. Ask your stakeholders: if we had to pick, what matters most? In some cases, the choice is already made for you: depending on your industry or location, accessibility and security might be legal requirements. For a bank, it's security. For a marketing website, maybe it's accessibility. The key is to get explicit about priorities. Just saying "Everything is important" is the same as saying "Nothing is prioritized."

CFRs also help shape technical decisions. Without them, you risk overengineering for imaginary problems. So many startups out there plan and build for millions of users when their business plan might support only 10,000, or even fewer.

Yes, business goals can change. If suddenly you do need to scale 10x faster, you'll deal with that when it happens. But it's better to adapt later than to waste months now building for problems you don't even have yet.

A very important CFR that comes with every system, and one we don't talk about enough, is cost. A simple trap teams fall into is overusing cloud resources. Cloud feels intangible: it's just so easy to spin up a new AWS EC2 machine, another database, another microservice. In the datacenter world, setting up a server meant real work, real approvals, real conversations with stakeholders. Now, it's just a couple of clicks. The easier it is to access resources, the harder they are to control. Just one more server. Just one more database. Just one more experiment.

Most teams realize the real cost too late, when the cloud bill spikes and reversing it is no longer easy. How fast you catch it often depends on the size of the company. In bigger companies, these things can go unnoticed longer. I've been there. I've had that tough conversation as a tech lead, explaining to my manager why our AWS bill was one thousand euros more because we forgot a server still running overnight after some experiments. Not easy.

The point is this: cost should always be at the back of your mind as a tech lead. Just because it's easy doesn't mean it's free. Cost is a real architectural concern, and you should treat it like one.

Good CFR management is about making things visible early, setting shared expectations, and helping the team make better trade-offs as they build. It's about working with what you know now, being intentional with your decisions, and adjusting when new information comes in.

In the end, thinking about CFRs early is just another way of doing the hard part of leadership: seeing around corners and protecting your team's future work from today's easy mistakes.

BALANCING INNOVATION AND STABILITY

The two extremes of innovation and stability often show up clearly in tech leads.

On one side, you have tech leads driven by trends, always chasing the latest tools, the newest patterns, the next big thing. They move fast, they innovate constantly, and sometimes they leave a trail of fires behind them. Their teams often become polyglot, using multiple programming languages and solving similar problems in different ways. Microservices multiply, each slightly different, and maintenance slowly becomes harder and harder over time.

On the other side, you have tech leads who stick to what they know. They push back on anything new, favoring stability and consistency. They reach for the familiar solution, even when it might not be the best fit, because it feels safer. Change makes them anxious, especially when it risks destabilizing the team or the product.

Finding a balance between these two extremes isn't easy, but it's necessary, and it usually comes with experience.

My suggestion is simple: get curious about your default style as a tech lead.

When you face a new technical challenge, what's your instinct?

Do you gravitate toward new tools and technologies, eager to try something modern, excited to bring fresh ideas into the codebase? Are you naturally drawn to innovation and exploration?

Or do you tend to lean on the tools you already know? Do you focus on stability, consistency, and making the most of what's already there? Are you naturally cautious about introducing unnecessary change?

Neither approach is inherently better. Both instincts can be valuable, and both can backfire if taken too far.

If you push too hard for constant innovation, you can easily overwhelm people, especially those with less experience, who need time to build depth. Your system grows more complex, as different problems get solved in different ways. Long-term maintainability suffers. And the faster you adopt new technology without a clear purpose, the easier it is to accumulate technical debt, inconsistencies, partial migrations, and abandoned experiments that will slow you down later.

If you cling too hard to stability, you can end up slowing the team's growth. You can demotivate ambitious engineers who want to learn and improve things. They'll feel blocked, and they'll leave.

I know this from personal experience. I naturally lean toward stability. As a new tech lead, I always pushed my team toward consistency. Every time a team member proposed a new library or tool, I felt a knot in my stomach, because I knew I'd probably have to say no. I was afraid of losing control: over our code, over our processes, over our security.

But over time, I built more trust in my team. I also listened carefully to their feedback. Some of them shared, gently but clearly, that my resistance to new ideas was limiting their growth and motivation.

So I started to change, slowly. Small experiments here and there. Nothing huge at first.

Then one day, a developer in my team proposed something bigger: running a machine learning experiment on one of our core features. He shared the idea with a few stakeholders and a few team members. Everyone was excited. I wanted to support it, but I also knew I couldn't guide him deeply through the technical side, because I didn't have enough experience myself.

Instead, I leaned into trust. We sat down together and built a plan. We agreed he would own the project end-to-end: implementation, communication with stakeholders, tracking progress, defining success metrics. He would be the face of the project, but we would still share responsibility for how it turned out. That meant I'd stay involved, giving him visibility and support, setting up a clear, continuous check-in process between the two of us, and stepping in directly only if needed.

It worked better than I ever imagined. He delivered on time. The results were great. He got the rest of the team excited, started knowledge-sharing sessions, and helped upskill everyone. The product improved, and we built a reputation inside the company as an innovative, forward-thinking team.

The lesson here isn't "innovation is always good" or "stability is always bad." The lesson is: you can expand your leadership style, one step at a time. You can find ways to experiment that feel safe, for you, for your team, for your stakeholders.

We're seeing the impact of this balance challenge everywhere today, especially with the rise of AI. Some tech teams are speeding ahead, experimenting with AI wherever they can. Others are taking a more cautious approach, concerned about quality and long-term implications, they prefer to carefully explore what AI can and cannot do. Both mindsets are valid, and finding the right pace for your team is key. For more on how to approach this shift in a thoughtful, sustainable way, see the section "Integrating AI into Your Team" on page 28, where I explore strategies to support your team through this transition.

Balancing innovation and stability is just another part of the everyday reality of being a tech lead, especially in a space as fast-moving as tech. The ability to continuously explore your leadership style, and adapt it not just to your preferences but to your team's and company's needs, is more critical than ever.

There is no fully right or wrong answer. Context is everything. Environment matters. The needs of today might not be the needs of tomorrow.

As long as you keep checking in, with yourself, with your team, with your stakeholders, and stay open to the idea that you might also be wrong sometimes, you'll stay on the right path.

Integration and Deployment

Integration and deployment are two of the most critical technical processes in any product development cycle. They are the bridge between the code your team writes and the value that code delivers to your users. Smooth, reliable integration and deployment processes improve engineering efficiency and build trust across your team, your stakeholders, and your customers.

In this section, I'll focus on how to think about and continuously improve your CI/CD practices, define and refine your path to production, and ensure that your testing supports the speed and safety your product needs to evolve.

DELIVERING VALUE CONTINUOUSLY

In the past couple of decades, we've seen several practices showcase how speed and reliability actually go hand in hand when delivering software. DevOps, extreme programming (XP), continuous integration (CI), and continuous delivery (CD) have all demonstrated that shorter iterations tend to improve the stability of production systems.

Shorter iterations reduce the size of changes in each release, encourage better communication across roles, encourage a culture of automation, and enable early and continuous feedback. All of that leads to improved software stability and quality. Continuous deployment (also CD) is the natural next step in this line of thinking.

Before diving in, let's clarify the terms:

Continuous integration

> Continuous integration means automatically building and testing each code change as it's integrated into the main branch. This ensures that integration issues are caught early and often, making it safer to move quickly.

Continuous delivery

> Continuous delivery builds on CI by ensuring that every change that passes automated tests is ready to be deployed to production. However, the actual deployment still requires manual approval; a person must push the final button.

Continuous deployment

> Continuous deployment goes one step further: every change that passes the pipeline is automatically deployed to production without any human intervention. There's no faster way to get code running in front of your users.

Note

While the abbreviation "CD" is used for both continuous delivery and continuous deployment, the difference lies in whether the deployment step is manual (delivery) or automatic (deployment).

You can absolutely do CI without CD, but not the other way around. Continuous delivery or continuous deployment can't happen without continuous integration. That's why you always see them bundled together as CI/CD.

I was doing continuous deployment with my team back when pushing every change directly to production still felt a little bit crazy. I thought it was awesome; I couldn't stop talking about it. In fact, my first-ever public speaking gig was about continuous deployment. That's also when I first saw the pushback firsthand. People would ask, "Only working on the master branch? Every commit going directly to production?"

And to be fair, if you're coming from a world of long feature branches, extended phases of manual testing, and formal approval processes, then yes, code going to production within minutes can be quite the culture shock. It definitely took me some time to get comfortable with it.

Even today, many teams treat CD as aspirational, something to aim for. Some are still doing fully manual deployments. And many developers are outright against it. I think it's because they've never truly tried it, or haven't seen it done well.

I get it. When I first joined Thoughtworks, I thought the idea of deploying code multiple times a day was insane. I came from an environment where we deployed once per sprint—maybe. Each release required manual approval from a manager, followed by a carefully orchestrated testing process. The idea that code could go to production automatically, minutes after being merged, felt reckless.

But once I experienced it done well, with proper testing, monitoring, and safety nets, there was no going back.

The thing about continuous deployment is that it's not just about tooling. It requires a fundamental shift in how the team works together. It really works only when the team collaborates effectively. It requires vulnerability, committing small changes daily, even when things aren't "done." It requires a strong code review process, whether through pair programming or timely, thoughtful pull request reviews, alongside reliable tests and fast feedback loops. It needs a shared definition of done and clear agreement on what "ready for prod" actually means. Everyone has to be aware that a broken commit can block the whole team.

It demands extra care: making sure code runs locally, all tests pass, and new features are tucked safely behind flags.

Personally, I believe this is how software products should be developed and deployed. It just makes so much sense: fast feedback, no blockers, and a workflow that forces collaboration by design.

Now, I do acknowledge there are exceptions. Some industries, like medical or legal, might require human sign-off for every deployment due to compliance or legal constraints. In those cases, someone might literally need to push a button.

Also, there are risks in implementing this approach. It's a big cultural and technical shift. Continuous deployment requires serious discipline: strong test coverage, reliable pipelines, feature flagging, and a team that treats every commit as production-ready. In complex, distributed systems, one bad change can ripple across services in unexpected ways.

So yes, fast delivery comes with responsibilities. And just to be clear, I'm not saying you need to embrace full continuous deployment. What I am saying is that as a tech lead, you should be aware of where the industry is heading. Continuous deployment represents the current state of the art in modern software delivery. You may not need it, you may not be ready for it, but understanding its principles and trade-offs is part of growing as a technical leader.

Tip

If you're looking for a hands-on guide to implementing continuous deployment, check out *Continuous Deployment* by Valentina Servile (O'Reilly). It's a practical, thoughtful deep dive into the what, why, and how of modern delivery practices.

DEFINING YOUR PATH TO PRODUCTION

Regardless of where you are in your CI/CD journey, your code has a path to production. Sometimes it is straightforward; most times it is very messy. Whatever the state, it's your responsibility as a tech lead to ensure smooth ways of working, including your deployment process. That means making sure it's fast, smooth, and reliable and that you're continuously looking for ways to improve it.

But you can't improve what you don't understand. Start by gaining clarity on your current deployment pipeline: What does it look like today? What's working and what's not?

To answer that, you need to visualize your team's path to production, a step-by-step map of how code (software change) moves from a developer's machine all the way to production. It should capture the steps, people, tools, tasks, and output involved in turning ideas into running software.

I still remember the first time I had to draw my own path to production on a board and explain it to other tech leads as part of a tech lead training. I realized how many gaps I had in the process: things I was not sure about regarding how the pieces were connected and worked together.

So I brought it back to my team, and we filled in the pieces together. We even identified a couple of small changes that could reduce our time to production significantly.

I do have to mention that we were an autonomous team running on the "you build it you run it" methodology: this means we were in charge of not just developing the product but also deploying, testing, and maintaining it. There was no other infrastructure or QA team doing this for us. The company did have a vertical SRE (site reliability engineering) team, but they were managing just the very low-level underlying infrastructure and resources, like setting up the actual machines in the datacenter or setting the direction for our cloud architecture. We did have to have constant communication with them, but the rest (deployment pipelines, resource management, infrastructure cost management) was on us.

I am saying this because I know this is not the case everywhere. There are tech teams that are somewhere in the middle: where infrastructure teams play a bigger role in this process, as they define clear guidelines on how to do it, and you, as a team, have to apply them, but still build and manage your pipeline, and you have a bit more playing space with the settings and resources.

And there are teams that completely rely on other teams, usually referred to as *infrastructure teams*, to deal with everything deployment-related: these infrastructure teams define the rules and enforce them. They build the deployment pipelines, and you, as a team, depend on them for any changes you might want to make in the process. This goes to the extent that some product teams don't even have access to production, just to in-between stages, and the infrastructure team is the only one that manages the production deployment.

They are all scenarios that you might fall into as a tech lead, and before deciding which one is better, remember that the particular context of the team and business and company has a lot to do with defining what is good. So, before making any rash decisions, try to understand what the current process is and why it is like that.

Here is how you can easily get started with visualizing the process.

The template shown in Table 8-1 helps you visualize your team's delivery pipeline by mapping out two key dimensions:

X-axis (horizontal): Steps in the process

Each column represents a distinct step your code goes through, from the moment it's written to when it reaches users. This could include things like reviews, testing, approvals, deployments—whatever makes up your specific delivery flow.

Feel free to rename or adjust steps based on your actual process.

Y-axis (vertical): Key dimensions for each step

Each row captures a different aspect of that step, such as:

- People involved: who's responsible or contributing here?

- Tools used: what systems or platforms are being used?

- Tasks performed: what actions are taken?

- Outcomes/outputs: what is produced or expected at this point?

Table 8-1. Path to production template

	Development	?	QA	?	Prod
People involved					
Tools					
Tasks					
Outcomes/outputs					

Note

You could consider adding an analysis column to the "path to production" template.

Earlier, I described the path to production as the journey a piece of code takes from a developer's machine to the production environment. That's the typical framing used by most engineering teams.

But depending on your context, it might be useful to go a step further and include the pre-development stages too. Consider adding a column for analysis, the point at which a story enters planning. This is often where work starts to move and where bottlenecks can emerge: waiting on UX input, legal reviews, or decisions from other teams. Mapping these steps can help you understand delays that happen before any code is written.

It's completely optional, but if your team often gets stuck early in the process, it might be worth including.

If you have multiple systems or tech stacks, you should have a path-to-production diagram for each. The good news is these diagrams shouldn't change too often, so once you have them, maintaining them is relatively easy.

The value comes from doing this as a team, with all roles involved. So make sure to bring your draft in front of your team as early as possible and fill the gaps in the process together by answering these questions together:

- What are your dependencies?

- What steps along the path to prod are still not clear to you as a team? Who can help you fill in the blanks?

- In the Development column, consider the following: Who usually participates in development on this team? If you are doing pair programming, what are the pairing rules?

- What branch strategy does the team use, trunk-based or feature branch? How is the continuous integration process of the team? Are there any rules for using Git? What happens when there are doubts about the scope of the task? And when new scenarios are discovered?

- In the QA column (sometimes called Sign-Off), consider the following: Who participates in this step? What happens if the story is found not to be implemented as expected? What happens if bugs are discovered during?

- Are there any pre-production environments that need to be mapped out? Do you have dedicated environments like Dev, Staging, or QA? When are they used and by whom? Are there environment-specific checks, access rules, or blockers that impact the path to production?

- In the Production column, consider this: What is the process to deploy to production? Any manual steps involved? Is there someone outside the team who participates in this? Who? What happens if the deployment is not successful?

There are endless benefits to visualizing your path to production. Here are just a few:

It makes your deployment process faster and smoother

Once you've built the diagram, sit down with your team and walk through it together. Ask the right questions: What are the bottlenecks in our process? Where do we have long wait times? Where does work tend to pile up?

From there, identify small improvements you can start working on today. Sometimes the smallest change can have the biggest impact. For example, simply reordering tests or upgrading a library might significantly reduce waiting times.

In one team, there were flaky automated tests causing random failures in our deployment pipeline. Every time a test failed, someone had to investigate, confirm it was a known flake, and manually restart the pipeline. Over time, this became so normalized that no one questioned it anymore. "Oh, a flaky test. Just rerun it." It was part of our daily routine.

These tests had been around forever. No one really knew what they covered anymore, and people were hesitant to remove them, just in case they were catching something important.

Once we started working on our path to production, identifying bottlenecks like this one, we also added metrics to see how often it was happening. Surprisingly, one out of five deployments was failing because of these flaky tests, way more than we thought. We decided to take action.

We reviewed all the tests that were frequently failing. Some we fixed. Others we removed, accepting the trade-off that if something truly critical broke, we'd catch it elsewhere. The truth was, these flaky tests were giving us a false sense of security. People had stopped trusting them, so in practice, they weren't protecting us at all; they were just slowing us down.

Once we cleaned them up, our pipeline became more stable and more trustworthy.

It was only when we mapped our path to production that the real impact became clear, and once we saw it, we couldn't ignore it anymore.

It improves stakeholder conversations

Becoming aware of your bottlenecks, delays, and improvement areas, and sharing them with stakeholders, can be a game-changer. This kind of transparency helps justify large estimates that depend on other teams, highlight blockers or handoffs, and build a shared understanding of why something is taking time. It also feeds into better forecasting and planning.

It strengthens collaboration

Mapping out your path to production can bring teams closer together.

On a new project, it helps define how you'll work as a team from day one. Rather than inheriting old habits or building a release process reactively, you can be intentional about the steps you take and how you collaborate.

On an existing project, it can reveal process gaps or mismatched expectations. A shaky path to production often reflects deeper communication or collaboration problems.

Beyond your immediate team, these diagrams can also lead to broader improvements at the company level. When you start conversations with other teams about shared bottlenecks or dependencies, you raise visibility, and sometimes trigger momentum for cross-team process change.

It helps with onboarding

Path-to-production visualizations make excellent onboarding tools. New joiners may not remember every step, but having a clear visual map gives them a solid overview and something to refer back to when they're unsure. I strongly believe every onboarding process should include walking through this diagram together.

The value of visualizing your path to production is just as important even if your team does not control most of the process. A common question that comes up is "What value is there in analyzing something we don't control?" It's a fair question.

But the most effective tech leads are the ones that are involved in all stages of development of their product, including how efficiently and reliably it makes it to production and whether it meets business expectations. Even if parts of the process are owned by other teams, understanding those black boxes, knowing who manages them, and being aware of the delays they introduce can help you work more effectively with those dependencies.

Take, for example, a common scenario: your deployment pipeline is managed by a site reliability engineering team. It might be tempting to say, "The pipeline is slow; that's their responsibility to fix." But it's often more productive to take shared ownership.

In one team I worked on, our pipeline suddenly started slowing down. After about a week of waiting for it to resolve, I reached out to the SRE team to ask what was going on and whether we could help. It turned out they'd recently made a change to a config file and hadn't realized it impacted our pipeline. Once they looked into it, they discovered the issue was affecting multiple teams, but since no one had raised it, they didn't know it was a problem.

By stepping in, we solved our issue but also helped the SRE team uncover a broader impact. That kind of awareness and collaboration happens only when teams stay curious about the full delivery process.

Ultimately, clarity on your path to production improves more than just delivery speed. It builds confidence, strengthens collaboration, and lays a solid foundation for consistently delivering quality work.

CONTINUOUSLY TESTING

These days, automated testing in software should come as a default requirement. The benefits are huge, and we all know it. But many of us have also felt the pain of untested code because, despite knowing better, the reality is that testing is often skipped. This is especially true in fast-moving areas like machine learning and AI, where excitement outpaces the maturity of the testing ecosystem.

But it's not just about new tech. Legacy code, the kind that holds systems together and generates most of the revenue, is often untested too. People are afraid to touch it, and for good reason: without tests, you never know what will break. That fear paralyzes improvement, blocks features, and raises the risk of bugs and outages.

Well-done testing brings trust. It brings safety. It tells you if something's broken before it hits production. It aligns the team on what the code should do. It documents what the code actually does. It gives you a history of how features have evolved. And most importantly, it builds trust with customers through stability and reliability.

Practices like continuous integration and continuous delivery depend on proper testing to be effective. Technically, you can ship without tests, but you'll likely face more bugs, more regressions, and slower recovery. Fast tests mean fast feedback, which means faster delivery.

A healthy testing strategy also impacts how your team collaborates. It affects pair programming, code reviews, and your ways of working. A team with no testing usually means people work in silos, and focus narrowly on getting their own code "done," without thinking about how it fits into the bigger picture or might impact others.

I could go on about the benefits of testing, but you probably already get the point. Let's focus on why this matters to you as a tech lead. You're accountable for everything your team delivers. That doesn't just mean "it works"; it means your product meets quality standards, is scalable, and is reliable. You simply can't achieve that without proper testing.

But what does "proper testing" even mean? That depends on your team.

In my experience, every team is different. I've worked on projects with zero tests, some with basic unit and integration tests, and others with full-blown strategies including TDD,[1] contract tests, end-to-end (E2E) tests, and automation across the board. There's no one-size-fits-all answer. I love TDD, but I also know context matters. Your team's environment, timeline, and maturity level all influence what makes sense.

Start by defining what proper testing means for your team. Begin by understanding the company-wide expectations. Are there minimum standards your team is expected to meet? Are there any tool restrictions in place that you need to respect? Once you have clarity on these aspects, assess whether your team is currently meeting them. Even if your team isn't aligned yet, as a tech lead, you need to be fully aware of these expectations because you are responsible for enforcing them.

While thinking about how to measure your testing efforts, be cautious with code coverage (a software testing metric that measures the percentage of source code executed during automated testing; in simpler terms, it tells you how much of your codebase is being covered by your tests). A lot of tech teams rely on coverage metrics to define how well their code is tested, but this can be misleading. I've seen flaky systems with 80% coverage and working systems with 20%. Code coverage doesn't tell the whole story. It doesn't account for quality, critical paths, or meaningful assertions.

If your company doesn't have clear testing guidelines, create them for your team. Sit down and build a strategy together. Start by identifying what types of testing are currently in place, whether that's contract testing, load testing, exploratory testing, or A/B testing.

Then, decide what testing layers you want to strengthen or introduce: unit tests, integration tests, E2E tests, and so on. Map out the tools you're already using, and note where your gaps are.

One helpful mental model to guide these decisions is the test pyramid (Figure 8-2). It emphasizes having more unit tests at the base, a moderate number of integration tests in the middle, and fewer E2E tests at the top. This structure helps balance speed, reliability, and coverage. Unit tests are fast and focused, integration tests check that components work together, and E2E tests simulate real user flows but tend to be slower and more brittle.

---

1 For a deeper dive, see Kent Beck's book *Test-Driven Development: By Example* (Addison-Wesley, 2002).

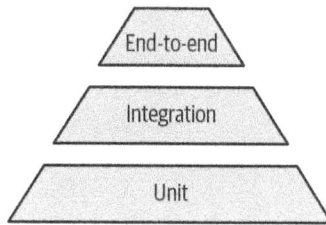

Figure 8-2. Test pyramid

This doesn't mean blindly chasing a perfect ratio, but it's a useful principle to avoid common pitfalls, like teams that rely too heavily on slow or flaky E2E tests without having a solid base of unit or integration tests.

Finally, clarify which parts of your testing process are automated and which are still manual. Having this full picture will make it easier to spot opportunities for improvement and build a realistic, actionable plan.

From here, shift your focus toward the biggest pain points standing in the way of better testing. Maybe tests don't run locally, and developers lose confidence before even pushing their changes. Maybe your CI pipeline is unstable, constantly breaking due to flaky tests that no one trusts anymore. Sometimes the tests are simply too slow, delaying feedback and slowing down your delivery process.

Other times, critical parts of your system, like edge cases or infrastructure components, might not be tested at all. And in some cases, ownership of testing might be unclear or fragmented between developers, internal QA, and external QA. Each of these issues contributes to inefficiencies and risk. Document it all. Whatever you decide to improve, now or later, having a written record will help.

Now that you've assessed your current state, the next step is to define what "better" looks like. This might align with broader company standards or focus on removing the day-to-day friction your team experiences. It could mean ensuring everyone can reliably run unit and integration tests locally, eliminating flaky tests that break the pipeline, or removing manual steps that delay deployments. Even small goals can lead to meaningful improvements when they're intentional and shared.

You don't need to fix everything at once. Start with the most painful, recurring issue that's relatively easy to address. For example, simply improving local test reliability can significantly boost team confidence and delivery speed.

Once you've identified what to improve, work with your team to define a clear, practical strategy. Agree on your goals and how you'll get there. Whether that means introducing new practices or refining existing ones, the key is to be deliberate.

Whatever your vision is, define it clearly and collaboratively with your team. That shared vision becomes your testing strategy, even if the strategy is to keep things as they are, as long as that decision is intentional.

Improving testing is a business win too. When you need stakeholder support for your testing improvements, connect it to outcomes they care about: faster delivery, reduced risk, and stronger customer trust. Frame improvements in business language, not just technical jargon. Help them see that better testing will help them move faster over the long term.

Once you have a clear view of your testing landscape, bring your team into the process. Collaborate on improvements. You may even be able to delegate ownership to someone interested in driving this forward, organizing team sessions, documenting the current state, and helping prioritize fixes. Remember: being accountable doesn't mean doing everything yourself.

Even when you delegate, your role as a tech lead remains critical. You do need to think deliberately about what testing should look like for your team. Stay close to the process. Even if a separate QA team handles much of the work, make sure you understand how testing impacts your timelines, your risks, and your delivery flow. Work closely with this other team. Stay alert to bottlenecks and blind spots.

Testing is one of the fundamental practices that underpins everything else we do as tech leads. It connects to how we manage risk, improve delivery speed, and build trust, both within our teams and with our stakeholders.

Common Technical Challenges and How to Overcome Them

Every tech team deals with messy, unglamorous problems. The kind that don't get talked about in architecture diagrams or sprint demos, but they slow you down, frustrate your team, and get in the way of delivering value. As a tech lead, you're expected to navigate these challenges and help your team do the same.

This section is about those common technical struggles that every team faces at some point: growing tech debt that never makes it into the roadmap, incidents that catch you off guard and leave everyone scrambling, and the constant question of how much documentation is enough.

I'll go through each of these with practical strategies you can use to tackle them—not perfectly but better. Because handling these well is what separates a team that survives from one that thrives.

MANAGING TECHNICAL DEBT

The only project without tech debt is the one that hasn't started.

Tech debt is part of building software; it's unavoidable. In fact, in my experience, the most successful, revenue-driving systems are often the ones that happen to accumulate the most tech debt. Why? Because they've grown fast to meet demand. Because they're actively used. And because there's no way to build a real-world product at speed without making some compromises along the way.

Sometimes, tech debt is intentional: you cut corners to move quickly. Maybe you duplicate code, couple components too tightly, or skip testing. You promise yourself you'll fix it later, when there's more time...but "later" rarely comes, as it's not properly prioritized. There's always another feature, another deadline that comes first.

Other times, debt creeps in even when you don't touch the code. Third-party libraries move forward, while your system stays still. Over time, what once worked fine becomes outdated, buggy, or even insecure.

And tech debt isn't just in the code. There's also product tech debt: all those old features nobody uses, built in hacky ways that no one fully understands anymore. I've seen this kind of debt slow teams down more than any messy codebase, because new work must still integrate with and tiptoe around those unknowns.

So no, you can't avoid tech debt. But you do need to manage it.

If you ignore it, it grows into a bottleneck. Every new feature takes longer. Outages become more frequent. Engineers grow frustrated and disengaged.

Beyond the technical consequences, unmanaged tech debt chips away at morale. When your team is constantly tripping over fragile systems and patching problems they didn't create, it wears them down. It becomes harder to feel proud of the work. And over time, that erodes retention and team health.

Fortunately, there are ways to keep tech debt from spiraling. The first step is making it visible. You can't manage what you can't see.

Start by identifying and listing all known tech debt items across your system. This process alone gives clarity on the scale and impact, and it helps frame future conversations with your team and stakeholders.

One practical way to do this is to create a "Tech Debt Wall." This could be a shared Confluence page, a board in Jira, or even a physical space in the office. The idea is to make the most pressing tech debt items visible and transparent to the entire team and relevant stakeholders. You can include context, impact, and links to related issues or documentation. This helps prioritize discussions and reinforces that tech debt is an ongoing part of your product's health.

Another effective method is using an effort-value matrix. This is a collaborative tool where your team evaluates each piece of technical debt by plotting it on a 2 × 2 grid, with "effort to resolve" on one axis and "value of resolving" on the other.

The goal is to move away from gut feelings and create a shared, visual representation of priorities. It helps uncover hidden quick wins (high value, low effort) and align team discussions around what really matters.

Plus, it's a great way to include product managers in the decision-making process, since it translates technical trade-offs into clear business value. Tasks in the top-left quadrant in Figure 8-3 (high value, low effort) are your low-hanging fruit and a great way to start.

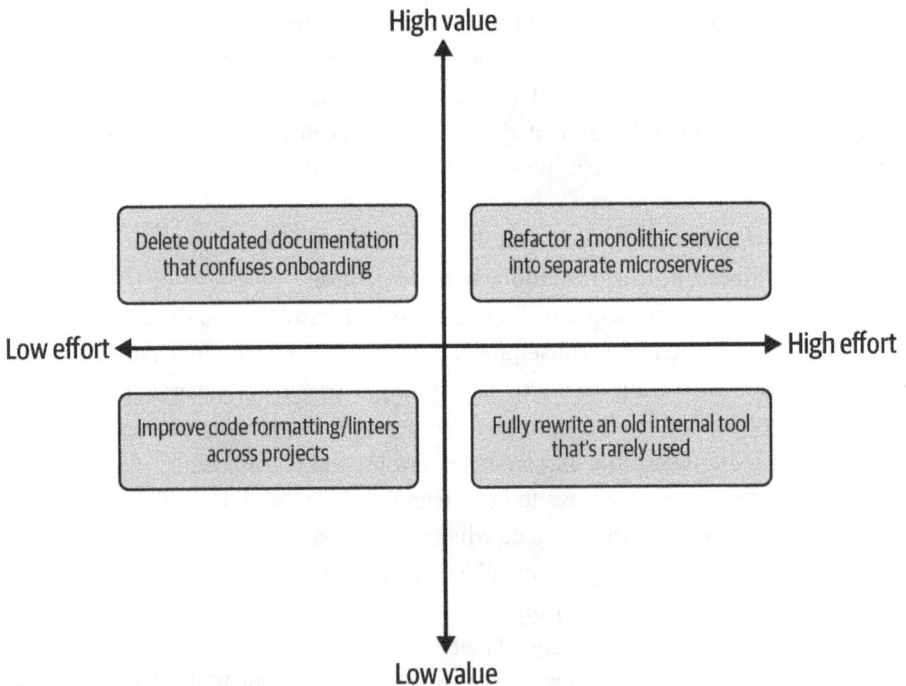

High value

Delete outdated documentation that confuses onboarding

Refactor a monolithic service into separate microservices

Low effort ←→ **High effort**

Improve code formatting/linters across projects

Fully rewrite an old internal tool that's rarely used

Low value

Figure 8-3. Effort-value matrix

In addition to the effort-value matrix, you can use more nuanced prioritization frameworks that take into account not just effort and value but also risk, frequency of impact, and alignment with strategic goals. For example, assigning scores to each of these dimensions, on a 1–5 scale, can help you generate a composite priority score. This makes trade-offs easier to justify and decisions more transparent when discussing priorities with stakeholders or leadership. Tools like RICE (reach, impact, confidence, effort) or custom-weighted models can be adapted for this purpose.

Use whatever framework fits your team's decision-making style, but the key is to move from opinion to evidence-based prioritization. Start somewhere. Prioritize the tasks. Add them to your backlog. Work with product people in your team to build a strategy around them.

Now that you have them prioritized, here are some strategies for how you can start introducing these tasks in your daily team work:

The 80/20 rule
Reserve 20% of your team's capacity each sprint to reduce tech debt.

Package tech debt work alongside relevant product initiatives
For example, if you're touching a part of the codebase to build a new feature, use that moment to clean up related debt in the same area.

Theme-based sprints
Focus each cycle on a particular type of technical debt, such as performance optimizations, test coverage, or security hardening. This helps the team dive deeper into a specific area.

Definition of done
Expand your DoD (definition of done) to include debt-related criteria (e.g., no new dependencies added without review).

In addition to visualization exercises, consider using tools that help you track and monitor technical debt over time. Platforms like Code Climate, SonarQube, and Stepsize that integrate directly into your codebase and CI pipeline, highlighting hotspots, code complexity, and maintainability issues.

For backlog-level management, tools like Jira, Asana, or Linear can be structured to tag and group technical debt items by type, impact, or urgency. The key is to choose tools your team will actually use and keep them tightly integrated into your workflows.

This sounds like a lot, but again, I have to remind you that as a tech lead, you need to ensure tech debt is taken care of. It's actually a great opportunity to delegate. In my team, we had a rotating role called the Tech Debt Champion.

This person's focus was to keep the list up-to-date, gather input from engineers, and surface the most painful areas of the codebase. You can even rotate this role, giving multiple people the chance to develop extra skills. You just have to ensure this person is properly set up for success to deal with the task. For a detailed plan on how to delegate effectively, check out Chapter 6.

When defining success metrics, avoid goals like "eliminate all tech debt." That's not realistic, or even necessary. Some tech debt is totally tolerable. In my team, we explicitly acknowledged certain debt we would live with, because we understood the trade-offs and risks. The key is intentionality: manage it deliberately, know why it's there, and make sure it doesn't silently grow into something harmful. The goal is sustainable control.

Tech debt is not the enemy. Poorly managed tech debt is. Done right, tech debt can actually be strategic. It lets you ship faster, validate ideas quickly, and seize market opportunities.

The next challenge is selling tech debt work to your stakeholders.

Let's say you've identified a few areas of tech debt that are slowing your team down. Maybe your CI pipeline is unstable due to flaky tests, a common testing-related form of tech debt that I explore in detail in the section "Continuously Testing" on page 245.

Or maybe you're dealing with outdated third-party libraries that are blocking new features, or a core module that hasn't been refactored in years and now takes hours to onboard new team members.

These issues are clear signs of tech debt, and they highlight the kind of day-to-day friction that creeps into development if left unchecked. You want to address them.

There are two ways to sell it to stakeholders.

The first is tactical and immediate: speak their language. Connect tech debt efforts to what they care most about: faster delivery, reduced risk, improved reliability. Explain how resolving flaky tests or refactoring brittle code can improve release stability, reduce hotfixes, and create space for faster feature work, fewer production issues, and happier customers. Start small. Pick something easy to fix that has a visible impact. Then measure and share the result: how much time was saved, how frustration went down, how the team feels more confident. Bring it up in retros, in sprint reviews, in casual updates. Let them see the improvement.

The second way is strategic and long term: play the long game. Build trust continuously. Don't wait until you need something to engage with stakeholders.

Keep them in the loop. Show steady progress. Invite feedback. Make them part of the journey. When you've built that relationship, getting buy-in is no longer a battle.

To navigate these conversations, use the three Rs of the stakeholder alignment framework:

Reframe

Translate tech debt into business value. Don't describe it as cleanup. Show how it leads to faster delivery, fewer incidents, and smoother onboarding.

Relate

Link the impact of tech debt to stakeholder pain points. Maybe it's been blocking a feature or increasing the time spent on support. Make it relevant to their goals.

Reinforce

Regularly share outcomes and updates. Build a track record of small wins to create confidence and momentum for larger investments.

I rarely got pushback from stakeholders as a tech lead because I put effort into building trust over time. I didn't talk to them only when I needed something; I kept them in the loop consistently, not just when things went wrong. I shared progress regularly, asked for their feedback, and included them in key moments. Over time, that steady communication built a foundation of trust that paid off when we needed space to focus on tech debt or internal improvements.

So when my team needed to tackle tech debt, there was not much debate: "We need to add this to the sprint. This requires us to remove this feature from the sprint, but we've already reviewed the timeline, and it won't have any negative impact." The response was simple and confident: "If you say it's needed, it's needed. Sounds good. I trust you."

The point of this story is to show you that there's a different approach, one where you collaborate with stakeholders instead of constantly pushing against them.

This kind of working relationship doesn't happen overnight. It requires openness on both sides, consistent effort, and a bit of good faith. But when you build that foundation, it changes everything. Conversations become easier. Buy-in becomes natural.

DEALING WITH INCIDENTS

Even in the best-engineered systems, failures happen. Servers crash. Deploys misfire. APIs time out. A third-party system breaks unexpectedly. These moments are stressful, but they're also inevitable.

There is nothing you can do to prevent all incidents, but there is a lot you can do to ensure your team is prepared to respond well when they happen.

Actually, how your team responds to an incident is a direct reflection of your engineering culture.

Build a culture of readiness and learning

Incident response isn't just reactive; it's proactive.

Start by normalizing incidents as a natural part of running software. Instead of being surprised or panicking when one happens, build a clear incident handling process that helps you stay in control. The following list will help you set up a solid, team-friendly response process:

Integrate company processes
> Many companies already have incident management processes at the organizational level. Make sure your team's process integrates with them. For example, some companies require different levels of stakeholder involvement depending on the incident's severity. Don't reinvent the wheel; adapt it and make sure your team knows it.

Define severity levels and classifications
> Not all incidents are created equal. Your login process breaking for all users is very different from it failing only for browser versions outside of your supported set. The number of users impacted, the risk of business loss: these factors matter. Your incident process should adapt based on severity: who gets involved, which communication channels are triggered, and how involved you need to be as a tech lead.

Make escalation paths obvious
> Everyone should know how to escalate an issue and where to find critical information like logs and metrics.

Invest in observability
> Go beyond alerting; make sure your systems are instrumented and your team can actually debug them (more on this in the section "Aim for Fast Feedback" on page 287).

Define clear on-call processes

You can't respond quickly to incidents if no one knows who's responsible.

That's why you need a clear and reliable on-call process, even for small teams. There should always be someone designated to keep an eye on alerts, investigate issues quickly, and coordinate a response when something breaks. Without this ownership, it's easy for problems to go unnoticed or for everyone to assume someone else is handling it.

Most tech teams solve this with an on-call rotation, where responsibility is shared across the team. Tools like PagerDuty or Opsgenie make this process smoother by automatically rotating who's on call, sending alerts, and even escalating if the first person doesn't respond. These tools help ensure no incident goes unacknowledged, whether it happens during working hours or in the middle of the night.

More on how to handle after-hours incidents in the section "How to deal with issues outside of working hours" on page 260.

But regardless of timing, the on-call process should be clear, fair, and automated.

Prepare for things going wrong

Consider running premortems: before major releases, gather your team and ask: "If something goes wrong, what would it be?" This simple question can reveal hidden risks, raise concerns that haven't been voiced, and help you stress-test your rollout plan. You'll often catch potential issues before they turn into real ones.

Alongside identifying risks, plan for your rollback path before you need it. It's not enough to hope you won't need one; make sure you can roll back safely and quickly. What's the trigger for deciding to revert? Who owns that decision? How long will it take? What's the impact?

Prepare for postmortems

Have a simple template ready to make postmortems easy to run and lessons easy to document. (More details to come in the section "What to do post-incident" on page 259.)

Train and practice

Make sure your team and stakeholders are familiar with the incident process. Best way? Run dry runs. Host incident drills, also known as game days, where you simulate real failure scenarios and go through the full response flow.

Even just one or two a quarter makes a massive difference in prepared-ness. These exercises help everyone stay sharp, ensure responsibilities are clear, and test the actual tools and processes you'll rely on during a real incident.

This kind of practice is especially valuable when things are calm. Processes that aren't exercised can quietly go stale, so it's good to take advantage of quieter times to review them.

You can also experiment with more advanced resilience practices, like chaos engineering tools (e.g., Chaos Monkey), to intentionally inject failure and test system and team responses.

The more familiar and practiced your team is with the process, the less stressful incidents will be, and the more likely you are to turn a crisis into an opportunity for learning.

What to do during the incident

Even with the best preparation, incidents will happen, and how you lead during them matters. When an incident strikes, your role as a tech lead becomes even more crucial.

First, make sure people are not losing it. During an incident, your team will look to you to set the tone. If you panic, they will panic. If you stay calm and steady, they will stay calm and steady. So take a breath, stay composed, and help them focus on solving the problem.

Second, make sure everyone is clear on their role during the incident. Clear roles remove chaos. You should hear things like "I'll check the logs," "I'll moni-tor alerts," "I'll coordinate with the PM and customer support." Most problems under pressure happen when people step on each other's toes or block each other because no one knows who's supposed to do what. Don't let that happen in your team.

Next, regardless of whether you are officially the incident owner or not, you need to protect the developers who are solving the problem. Shield them from stakeholder noise. Very importantly, don't put more pressure on them. Support them. For example, if a senior leader starts pinging the developer for live updates during an incident, step in and offer to provide updates yourself so the team can stay focused on resolving the issue. And if you're not the assigned owner, jump in to help manage stakeholders if you see that the owner is struggling or overwhelmed.

Another key part of your role is removing blockers. If the solvers need anything, like access to a certain system or support from another team, you step in. Use your authority and network to speed things up so they can stay focused on fixing the issue.

Now that we've covered your specific leadership role during an incident, let's shift into the practical side: how to manage the incident process itself:

Assign a clear owner

Every incident needs a coordinator: someone who keeps the response on track by organizing efforts, maintaining communication, and shielding the team from distractions. This person is responsible for sharing updates with the team, stakeholders, external teams like SRE, and users if needed. After the incident, the coordinator also leads the postmortem, assesses business impact, coordinates follow-up work, and ensures documentation is updated.

In many teams, this role naturally falls to the on-call engineer, which keeps the process straightforward.

Tech leads often feel the urge to step in and take on this responsibility by default. But in practice, the process runs more smoothly when others are empowered to lead, and the tech lead supports them while focusing on broader priorities like business impact and long-term strategy.

Define the issue and assess severity

The first step is helping the owner define the severity of the incident. Before jumping into solution mode, they need clarity on what's actually happening and how serious it is.

Sometimes the impact is obvious: your system is down and users are blocked. But other times, like in edge deployments or customer escalations, the scope is less clear.

Support them by bringing in the right people: engineers, product managers, support staff, and business stakeholders. Together, collect the facts and answer: What exactly is happening, and how widespread is it? Are users impacted? Is revenue, data, or reputation at risk?

If your company has a documented severity framework, this is the time to use it. If not, add it to your post-incident list to define one. Clear thresholds prevent noncritical issues from shortcutting regular delivery processes and being treated as urgent unnecessarily.

Starting with severity ensures the team responds with the right level of urgency and keeps things proportionate, focused, and aligned.

Clearly define solvers

The people actively working to resolve the problem must be explicitly identified. Without clear roles, everyone scrambles and progress slows. Solvers should be protected from stakeholder pressure and distractions.

It's also worth emphasizing this: the team's top priority during an incident is to restore service, not to find the root cause. That can come later. Often, a simple revert or a temporary scale-up can stabilize things quickly. Once the system is back in a healthy state, you can take your time identifying what went wrong.

Establish a clear communication process

Poor communication can turn a manageable incident into chaos. If there are no regular updates, people get nervous. If solvers are constantly interrupted for status checks, they can't focus.

Agree up front how you'll communicate: regular verbal check-ins, updating a shared doc, or quick syncs every few hours. Personally, I prefer quick live check-ins to make sure everyone's aligned. Whatever you pick, stick to it. Resist the urge to start pinging solvers; your role is to act as a shield so they can focus on fixing the issue.

Stakeholders need updates, even when there's no major progress yet. If you don't update them, anxiety rises, and they might start distracting the team. Establish early on who communicates with them, how often, and through which channel, whether it's a Slack thread, email updates, or short status meetings.

If the incident affects users, you may need to send updates through Statuspage, email notifications, or other public channels. Even when you don't have all the answers yet, sharing early updates helps build trust and shows transparency.

Track progress visibly

Log follow-ups clearly. Use Jira, Confluence, your tech debt board, whatever works. Just make sure it's visible and owned.

With the right leadership and preparation, incidents don't have to be disasters; they can be the moments when your team becomes even stronger.

What to do post-incident

Once the incident is resolved, the real work begins.

First, make sure any official documentation that needs updating is taken care of: business impact reports, incident summaries, customer communications, anything stakeholders need visibility on. This should be the incident owner's responsibility, but as a tech lead, you should make sure it happens.

Next, create runbooks for common issues. If something caused an incident, document it. Make it easier for the team to respond faster and better next time.

Every incident is a learning opportunity, so treat it like one.

Run a blameless postmortem within 48 hours while the details are still fresh, regardless of how "big" the incident was. Invite everyone who was involved: the main team, supporting teams, and relevant stakeholders.

Some teams even go a step further and make their incident review meetings open to all engineers in the company. Anyone can join, listen in, and learn. This level of transparency isn't required, but it might be something you want to consider.

The purpose of a postmortem is to understand what went wrong in the system and identify how to improve it. The key questions to answer are: What happened? Why did it happen? What can we improve? Some of the best improvements come from incidents.

Documentation is a crucial part of turning incidents into learning opportunities. A well-structured postmortem makes it easier for others to learn from the experience and ensures transparency. A good postmortem usually includes the information reflected in Table 8-2.

This structure ensures clarity, transparency, and shared learning across the team and organization.

Share the documentation widely, and make sure the whole team (and even other teams) can learn from it.

Make sure to follow up on action items. Assigning tasks isn't enough. Set a clear due date for each one, and check in a few weeks later to make sure follow-ups are completed.

Add them to the team's regular workflow, whether that's Jira, Trello, or your shared Kanban board, so they stay visible. Make sure they are prioritized alongside other work and given the right level of urgency.

Without this kind of accountability, postmortem learnings stay talk instead of action.

Postmortems help build a culture where it's safe to talk about problems, and safe to get better, together.

Table 8-2. Postmortem template

Postmortem template	
Date	Date of the postmortem
Author(s)	People documenting the postmortem
Status	Assuming the incident is resolved, here, you can add extra details about the status, like "action items in progress"
Summary	A quick overview of what happened
Impact	What was affected and how
Priority	The severity classification (e.g., low, medium, high)
Incident coordinator	Who was responsible for managing the response
Start time	When the incident began
Resolved time	When the immediate issue was resolved
Closed time	When all follow-ups and documentation were completed
People involved	List of everyone involved in the response
Incident communication channels	Slack channel or comms method used for coordinating the response
Customer communication	Whether customers were affected and if communication was needed
Root cause(s)	What triggered the incident
Detection	How it was discovered or missed
Response	What actions were taken
Timeline	Key events and timestamps
Lessons learned	What went well, what didn't
Follow-up actions	Specific tasks with owners and deadlines

How to deal with issues outside of working hours

I used to be completely against the idea of on-call support. My mindset was always "We need to build systems that don't require people to be woken up at night." I thought my team was fully aligned with that too. So when the client suggested introducing an official paid on-call process, I assumed it would be a hard no from us.

Surprisingly, some team members were interested. Because the reality is, no matter how good your system is, there are always things that can go wrong outside working hours. And the truth was, we were already covering incidents outside of working hours, just informally, and we weren't getting compensated for it.

In the end, we voted as a team. We decided to stick to no official on-call system, but we promised to work even harder to minimize after-hours issues. We also introduced a small reward system: if you had to work at night, you could take time off during the day, one hour for every hour worked. Looking back, I realize now that pushing so hard against on-call might not have been the best move for my team.

The reality is that systems will break outside working hours. Some problems can wait until morning or Monday. Others can't. Either way, you need to have a clear conversation with your team and stakeholders about how you'll handle these situations. And most importantly, whatever system you put in place, make sure people are properly compensated for their time and effort.

I've seen two common models for on-call compensation:

Per time

> You get paid for simply being on call during a rotation window, whether or not anything breaks. I've noticed that this model motivates teams to improve their systems, accumulate less tech debt, and have fewer recurring issues, because there's a better chance of "getting paid to do nothing."

Per incident

> You get paid only if you actually have to jump in and fix something. This can sometimes backfire; if people get paid only when things go wrong, there's less incentive to proactively fix root causes or invest in system improvements.

Whatever you choose, sit down with your team and stakeholders. Define clear on-call rotations, responsibilities, and expectations. Document the process, make it visible, and add it to your onboarding checklist and ways of working as a team.

A great incident response process is a team effort. It starts with preparation, shows up in how you communicate during a crisis, and continues long after the issue is resolved. When handled well, incidents can actually strengthen your systems, your team, and your trust with stakeholders.

DOCUMENTATION OR NO DOCUMENTATION?

There's always been a long-standing divide among developers when it comes to documentation. Some are firmly against it, while others are fully in favor.

As always, the answer is somewhere in the middle, and it depends on what kind of documentation we're actually talking about.

The common critique is that documentation goes stale and that the code should be the ultimate source of truth. It's true that some forms of documentation, like implementation notes, can become outdated quickly, and yes, in an ideal world, code would be expressive and clear enough to speak for itself. But let's be honest: how often is the codebase that clean and easy to understand?

On top of that, many things simply can't be captured in code. Team processes, roadmaps, architectural decisions, operational runbooks: these aren't things you can read from a function or a class. These kinds of documents don't exist to explain what the code is doing but to explain why decisions were made and how to work with the system and the team behind it.

I don't think the right question is "Documentation or no documentation?" It's more about how to do documentation effectively. What should be documented, and how do we keep it useful?

Personally, I lean toward writing things down. As a tech lead, I've seen how valuable even lightweight documentation can be, especially for onboarding new team members, reducing dependencies on specific individuals, aligning the team, and debugging issues that surface months later.

Yes, documentation takes effort, and no, it won't always stay perfectly in sync with reality. But even slightly outdated docs can still be useful; they give you something to update, challenge, or validate. That's far better than having no starting point at all.

The value of documentation

Documentation plays a powerful role across many aspects of your work as a tech lead. It becomes critical when tracking architecture changes (check out the section "Visualizing a System Architecture" on page 226), managing tech debt (check out the section "Managing Technical Debt" on page 249, and learning from incidents (check out the section "What to do post-incident" on page 259). Without clear records of why decisions were made or what trade-offs were accepted, teams fall into confusion, repeat mistakes, or rebuild solutions to problems they already solved.

The biggest value of documentation is alignment, with your team, stakeholders, and even your future self. It keeps reasoning visible and decisions anchored, and it gives everyone a shared reference point when things drift: "This is what we agreed on; what changed?"

A great example is documenting your ways of working. These are rarely written down, because everyone assumes: "We know how we do things; we do them every day." But that might backfire.

I once coached a tech lead who was frustrated by shifting priorities. The CTO kept assigning engineers urgent tasks directly, bypassing the team's sprint planning. "She should follow our process," he said.

So I asked, "How would the CTO even know your process? Have you walked her through it? Is it written down anywhere?" He paused: "No...I guess I just assumed she knew."

So he wrote a short, high-level summary of how work enters the backlog, how it's refined, and how prioritization is handled. Then he shared it in the team's workspace and tagged the relevant managers, including the CTO.

Two days later, the PM messaged him: "I've added a task from the CTO into the backlog for the team to evaluate. We can discuss in refinement." No more random interruptions. The process held.

This example highlights more than just a documentation gap, but the takeaway is simple: sometimes, all it takes is writing things down and making them visible.

Also, documenting your ways of working can be incredibly valuable for onboarding. New joiners often struggle with unspoken norms or informal processes. Having these written down—how planning works, how priorities are set, what "done" means—gives them something concrete to refer to.

But it's not just new joiners who benefit. Even experienced team members feel the impact when key knowledge is missing. Especially when systems break—and they will—good documentation can save your team hours of confusion and stress.

One client shared this story with me: her team was working with a legacy system that had become a major source of friction. Every time something broke, it triggered a firefight: hours lost digging through old tickets, chasing down people who'd once worked on the system, and trying to reverse-engineer what the code was supposed to do. Even small incidents caused big delays.

So she introduced a lightweight habit: any time someone figured something out about how the system worked, like how services interacted, how to restart a flaky job, or what a mysterious config did, they wrote it down in a shared runbook—not overly detailed, just enough for someone else to pick it up later.

Over time, that runbook became one of the team's most valuable assets. During one major incident, the team even used the runbook in real time to guide their mitigation steps.

It was a great reminder that documentation doesn't need to be big or formal to make a big impact. It can be as simple as writing things down as you go.

The great news is that these days, the documentation process is way easier because you have better tools. Screen recording tools like Loom, Scribe, or Tango can capture workflows in seconds. AI-assisted notes can summarize your actions and generate documentation as you work. Lightweight templates for runbooks, onboarding guides, and decision logs make it faster to create useful content. And if you update documentation alongside your pull requests and technical changes, it becomes a natural part of development rather than a separate "documentation day" months later.

Of course, there's a risk with these tools too. While they make it easier to generate documentation, they can also lead to disengagement, such as auto-generated docs that no one reads, questions, or updates. To avoid this trap, someone on the team still needs to take ownership: reviewing content, validating its usefulness, and making sure it stays relevant and integrated into your day-to-day work.

What to document

Here are some things worth documenting and keeping up-to-date as part of your team's day-to-day work:

Architecture
> Document the current state, desired state, stages of changes, and decisions made along the way (more on this in the section "Architectural Strategies and Implementation" on page 221).

Key decisions
> Why they were made and what alternatives were considered. This applies not just to architecture but also to product and team processes (more on this in the section "Track Technical Decisions—Architecture Decision Records" on page 281).

Incident postmortems and learnings

Capturing what happened, why it happened, and how to prevent it in the future (more on this in the section "Dealing with Incidents" on page 254).

Strategies

Examples include your testing strategies, your path to production (more on this in the section "Defining Your Path to Production" on page 239), objectives and key results (OKRs), and the contracts and agreements between your team and other teams or systems.

Team processes

Onboarding guides, ways of working, and anything that helps new and existing team members move faster and stay aligned.

Infrastructure

If you're using infrastructure-as-code tools like Terraform, your infrastructure can become self-documenting to some degree.

The goal is to document the parts of your system, culture, and processes that help your team move quickly and stay aligned, focusing on what's most useful and relevant to your daily work.

How to keep documentation alive

As mentioned before, the biggest issue with documentation is keeping it updated and relevant. There are a couple of strategies that can help you do that.

Make updating relevant documentation part of the "definition of done" for work items. This way, changes in architecture, processes, or systems are reflected immediately.

Assign a rotating "doc champion" role within your team to periodically review and flag stale documents. Fresh eyes help catch things that otherwise get forgotten.

Include documentation reviews as part of your retrospectives or team health checks. If something important changed recently, there's probably a document that needs updating too.

Continuously delete or archive old, misleading, or irrelevant documentation. I have to say this is one of the most satisfying parts of the role: the ability to just go and delete something that's no longer useful. Honestly, it's better than creating documentation. I know it can feel scary to delete things, so here's a strategy I use: archive before you delete. Move the document somewhere inaccessible but retrievable. See if anyone notices, complains, or needs it. If after a while, no one asks for it, you can confidently delete it.

Another approach I've used is simply removing access; if someone actually needs it, they'll request it. That becomes a great signal that the document is still in use. You can even take it one step further; when someone does ask for access, start a quick conversation: "What do you use it for?" It might turn out they need something more current or that they were referencing it out of habit, not usefulness. Worst-case scenario, if you do delete it, but it turns out you need it, you'll rebuild it, this time actually relevant and updated.

If keeping your documentation alive isn't part of your team's routine, it will rot, no matter how beautifully it started.

Documentation needs to be useful. Document for people, not for process. Make it easy, make it actionable, and make it part of how your team works every day.

Key Takeaway

Great technical leadership means guiding architecture, deployment, and every-day decisions in ways that help your team move fast and build with confidence.

Throughout this chapter, we looked at the real, practical responsibilities that shape a healthy engineering culture: evolving architecture, improving delivery pipelines, managing tech debt, handling incidents, and making documentation useful.

All of these are part of the same goal: making intentional technical choices that support your team and product over time.

Technical work is never separate from people work.

9

Managing Technical Projects

No matter how strong your technical skills are, stepping into a tech lead role means entering a new layer of complexity, one where communication, prioritization, and expectation management matter just as much as architecture and code.

From my experience, your biggest challenges won't be technical. They'll come from aligning people, managing expectations, and keeping your team on track when things inevitably shift.

In this chapter, I'll share practical ways to lead through that complexity. I start with how to define a clear technical vision and shape a strategy to achieve it. Then, I'll cover how to help your team make thoughtful technical decisions that align with that vision. Finally, I'll walk through how to track key decisions over time using tools like architecture decision records (ADRs).

Next, I focus on how to encourage technical excellence inside your team. I'll talk about what it means to define team standards, how to build fast feedback loops into your delivery pipeline, how to measure what matters, and how to ensure quality through meaningful, useful testing practices.

Finally, I'll walk through four of the most common project management challenges I've seen tech leads face: planning and keeping a project on track, dealing with delays when the plan goes off course, balancing multiple streams of work at once, and identifying and managing technical risks before they escalate.

This chapter is about leading with clarity. You won't always get everything right, but the more proactive, transparent, and collaborative you are in how you lead projects, the more trust you'll build, and the stronger your team's delivery will become.

Aligning Your Team on a Common Tech Strategy

Most problems in tech teams come from people not being properly aligned. Even worse is when everyone thinks they're aligned, but they actually aren't. That's how you end up with a feature that takes a month to build and turns out completely different from what the business expected.

I've seen it happen so many times.

You might be thinking, "But didn't they talk along the way? Didn't they have check-ins, showcases?" Well, yes. Communication happened, but everyone was so sure they were on the same page that they missed the red flags, until the product was in front of stakeholders, and it was obvious it wasn't what they needed.

Misalignment always comes down to poor communication between the people involved. And bad communication doesn't just happen at the start; it follows through the whole process. It's usually only when the damage is clear that people realize something went wrong.

Clear alignment means everyone knows why you're building something, has a shared high-level understanding of how, and has agreed on a rough timeline. Clear, written alignment brings commitment.

This is where your role as a tech lead becomes crucial. You're standing at the intersection between your team and the business, responsible for keeping everyone aligned and committed. Your team will be making a lot of decisions, technical and not, and you'll be right in the middle of that process, ensuring that both sides understand each other clearly.

To do this well, you need enough technical depth to guide your team's work, and the communication skills to translate risks, problems, and strategies back to your stakeholders, whether they're deeply technical or not.

In this section, I'll focus on the skills you need to develop and practical strategies you can use to help your team make technical decisions smoothly, take decision making to the next level by building a shared technical vision, and document decisions clearly to ensure long-term commitment.

Even though the examples will focus on technical decisions, the same strategies easily apply to any type of decision your team needs to make.

BUILD A TECHNICAL VISION (AND TURN IT INTO A STRATEGY)

As a tech lead, one of your most impactful responsibilities is helping your team define a clear technical vision, then shaping a strategy that turns that vision into reality.

A technical vision describes what "good" looks like for your team's technology. It's aspirational and directional. It defines your north star: the destination you're heading toward.

Example: "We want a resilient and maintainable architecture that allows any developer to safely deploy to production in under 10 minutes."

A technical strategy, on the other hand, is the plan for how you'll get there. It outlines concrete steps and trade-offs based on where you're starting, the resources you have, and the constraints you're working with.

Example: "We'll break our monolith into services starting with billing, adopt feature flags for safer deployments, and implement observability dashboards to surface errors faster."

A helpful analogy here is building a house:

- The business strategy is why you're building it. Maybe it's to house a growing family, create a coworking space, or sell it as an investment. It sets the purpose and defines success.

- The technical vision is the blueprint of the finished house: the overall design and feel. It shows that you want a two-story home with lots of natural light, an open kitchen, energy-efficient systems, and room to expand. It captures what "good" looks like when it's done.

- The technical strategy is the construction plan: how you'll build it, in what order, and with what constraints. Maybe you'll use modular construction to speed things up, reuse the foundation of an older building, or phase the work so you finish the essential rooms first. It reflects how you'll get there, given your tools, timeline, and budget.

- Tactics are the daily site decisions: what to fix today, which materials to use for the flooring, or how to respond when it rains and delays the roof.

Vision gives you the why and where. Strategy gives you the how. Both are essential. Without vision, you risk building in circles. Without strategy, you may never get to your destination, or go wildly over budget.

Before starting to work toward a direction, you need to assess your current context:

Business goals
What's the product aiming to achieve?

Team strengths and skills
What technologies are they confident in?

Cross-functional requirements
Performance, observability, security?

Team shape
What complexity can your current size and roles support?

Tech landscape
What are you already using? What's working?

There's a concept I find useful here: enabling constraints. Your vision and strategy have to live within the bounds of your company's reality, like a requirement to use Ruby on Rails or stick with monolithic deployment. Ignoring these constraints leads to frustration when decisions made by the team turn out to be impossible due to external limitations.

Instead, make these constraints visible early. Set the boundaries clearly so the vision is grounded, realistic, and actionable.

For both vision and strategy, visualization is required as it transforms these from an idea to something tangible you can work with.

Both can be a simple document or a visual board (e.g., a Mural board) containing different artifacts like the following:

- Architecture diagrams; you can use the C4 model (more in the section "Visualizing a System Architecture" on page 226)

- Path to production (more in the section "Defining Your Path to Production" on page 239)

- Cross-functional requirements (more in the section "Defining and Managing Cross-Functional Requirements" on page 232)

- Test or quality strategy documents (more in the section "Continuously Testing" on page 245)
- Ways of working processes (more in the section "How to Build a Healthy Team Culture" on page 185)

Not every team needs all of these; start with what's most relevant for your context. For example, you might use architecture diagrams to illustrate both your vision (the future state) and your strategy (the incremental stages to get there).

Tip

If you're interested in a deeper breakdown of technical vision versus strategy, examples, and more resources on how to build them, I highly recommend the chapter "What's a Vision? What's a Strategy?" in *The Staff Engineer's Path* by Tanya Reilly (O'Reilly).

When you've articulated both the why/where (vision) and the how (strategy), execution becomes a lot smoother. To bring the strategy to life, you'll want to do the following:

Break down initiatives into clear milestones
These should represent visible progress: things you can demo, ship, or measure.

Connect strategy to the backlog
If your vision includes improving reliability, that should translate into specific tasks: refactoring flaky tests, improving alerting, or addressing known error patterns.

Create feedback loops
Set regular checkpoints, like weekly syncs, quarterly reviews, or retrospectives, to assess whether you're on track, what's been learned, and what needs to change.

Track and communicate progress
Share updates regularly. Keep stakeholders aligned, and make it easy for the team to know what's moving forward, what's blocked, and where help is needed. Don't just focus on what's done; highlight the impact and tie it back to the vision.

Celebrate alignment in action

When a team makes a decision that reflects the technical vision, like choosing a simpler solution because it aligns with the goal of maintainability, call it out.

Conclusion: Vision gives you direction. Strategy gives you a path. Execution turns ideas into impact.

HELP YOUR TEAM MAKE TECHNICAL DECISIONS

Once you have agreed on a plan with your team, the next challenge is making daily technical decisions that move you toward that direction. As a tech lead, you'll be involved in decision making constantly.

Whether it's choosing between libraries, deciding how to integrate a new feature, or debating between building in-house versus buying off the shelf, your role is to support the team in making thoughtful, informed, and aligned decisions. You're there to guide the process, facilitate the conversations, and help the team weigh trade-offs.

The first thing is to make sure your team is clear on how decisions are made. Having an open conversation about your decision-making process can eliminate wrong assumptions. Many people are surprised when I tell them I won't make all decisions, that I want and expect them to contribute.

There are different decision-making styles out there. The most common ones are autocratic, consultative, democratic, and consensus (explained in Table 9-1). Each approach has its pros and cons and works better for certain types of decisions.

A note on consensus: consensus doesn't mean everyone has to agree 100% or love the decision. In practice, it often means everyone is willing to support the outcome and move forward, even if it's not their personal preference. This is sometimes called *consent*. People may not feel strongly in favor, but they don't strongly object either. Clarifying this up front can help avoid endless debates and make consensus-based decisions more effective in practice.

Table 9-1. Decision-making styles

	Style	What it is	When to use it	Example scenario
Leader driven	Autocratic	Tech lead makes the decision alone	Small, low-risk decisions or deadlocks when alignment cannot be reached and time pressure is high	Choosing a small library quickly Breaking a team deadlock near a release deadline
	Consultative	Tech lead gathers input, then decides	Medium-sized decisions that need some input but where quick resolution is important	Choosing between two cloud providers
	Democratic	Team votes and majority wins	Noncritical decisions where many opinions are equally valid	Naming a new internal tool
Group driven	Consensus	Team discusses until general agreement is reached	Big, strategic decisions where buy-in and alignment are critical	Defining the architecture for a new product

All of these decision-making styles are valid options based on the problem at hand. Problems start when you overuse one approach because it's more comfortable. My invitation here is: notice your tendency. Are you stuck using just one method?

Starting tech leads often fall into the trap of making all the decisions themselves. Maybe they feel pressure to know everything or think it's their job because they're accountable. But this creates bottlenecks, slows down the team, and reduces people's growth, including your own.

At the other extreme, when I started leading, I tried too hard to involve everyone. Every decision became a team discussion. I wanted to avoid being the leader who made all decisions alone, but in doing that, I slowed us down and frustrated my team. It took forever to reach consensus.

And then, under pressure, when I finally made a unilateral decision, people were relieved. That taught me: the problem isn't the decision style. The problem is staying stuck at one extreme. Balance comes from knowing when to involve everyone and when to make the call yourself.

The main thing that would have helped? Setting expectations up front. Let your team know how you plan to make decisions based on the context, whether you'll consult, involve, or decide quickly under time pressure.

This isn't rare. I was not the only one falling into this trap. I still see and work with tech leads every day who struggle with finding the right balance between making decisions themselves and involving their team. Just make sure you're using all the tools the role gives you. Build awareness of the options you have.

In general, you should aim toward group-oriented decision making whenever possible. Involving people leads to higher commitment and better results. And the most common way we have these days to make decisions as a group is through meetings. The bigger the decision, the bigger the meeting (more people involved). To make these decision-making meetings effective, you need strong facilitation skills.

Table 9-2 is a practical guide to the most common meeting pitfalls and what you can do about them in the role of facilitator.

Table 9-2. Common meeting pitfalls and how to address them

Common problem	Why it's a problem	How to address it
No clear goal	• Meeting feels pointless. • People don't know what's expected. • Time is wasted.	• Name meetings clearly ("Define delivery strategy for X" instead of "Talk about X"). • Add a description with meeting goals and links. • Start the meeting by stating the goal.
Meeting keeps derailing	• Conversation goes off-topic. • Important discussions are missed. • Meeting feels chaotic.	• Anchor the conversation: "How is this helping us reach our goal?", "Is this relevant for the conversation?" • Use whiteboarding (e.g., FigJam) to track the discussion live. • Split big problems into smaller pieces if needed. • Use RAID board to capture risks, assumptions, issues, dependencies.
Loud voices dominate	Only a few opinions are heard, and valuable ideas from quieter team members are missed	• "Pass the microphone": invite quieter voices to speak. • Use different tools to gather inputs: besides everyone sharing their idea out loud, you can use tools like anonymous forms, one-on-ones, or even write it down on Post-its. • Remote: make considerate use of the hands-up and chat features in your video conferencing tool.
Constant interruptions	• Hard to get your point across. • Conversations lose focus and energy. • People interrupted get demotivated and less engaged as they don't feel listened to.	• If you are interrupted, politely but firmly say, "I'm not finished speaking," "as I was saying before." • If someone else is interrupted, as a facilitator you have the power to bring that interrupted voice back: "Maria, I think you were saying something on this topic before?" • Create a rule to speak: raise your hand and ensure all raised hands are listened to.

Common problem	Why it's a problem	How to address it
Meeting drags on too long	Rushed decisions and fatigue	• Assign a timekeeper. • Split the meeting into timed sections. • Do time checks ("15 minutes left. Do you want to continue on this topic?"). • Be ready to reschedule if needed.
Meeting feels dead	• No one is engaged, and ideas stall. • Time wasted as the meeting is going nowhere.	• Pause and reassess if the meeting should continue. • Diagnose root causes: low energy, unclear purpose, lack of safety? • Adapt the format or split into smaller problems.
Meetings turn into fights	• Kills trust. • Derails teamwork. • Decision making breaks down.	• Bring people back to the problem you are trying to solve. • Shift from "me versus you" to "us versus the problem." Find common ground early. • Ask for clarification openly: "Can you explain that again?"
No clear outcome at the end	• Participants leave confused. • No clear next steps so nothing happens.	• Assign a note-taker other than you as facilitator. • Summarize what was discussed and what was agreed on, and assign next steps and owners. Validate assumptions live: "If I understand correctly, you mean X." • Write final agreements on a shared document live during the meeting.
Nothing happens after the meeting	• Wasted time discussing. • Erodes trust.	Have a clear process for following up: e.g., a quick follow-up message to the action owners, a shared doc to track progress, or a check-in meeting if needed.

To run effective decision-making meetings, you need more than a calendar invite. Here are some concrete steps to help you create structure, keep discussions focused, and ensure clear outcomes:

Set up a clear goal and structure

Before the meeting, ask yourself: if there's just one thing I want to achieve by the end of this meeting, what would it be? Not five things, just one.

Define it clearly in the meeting invite title and description. Add links to previous discussions or draft plans if they exist. When the meeting starts, restate the goal clearly and walk through a brief agenda, setting expectations for how the time will be used.

During the meeting, keep that goal top of mind. Whenever conversations start drifting, gently anchor people back by asking: "Is this relevant for the conversation?" or "How is this helping us reach our goal?"

Sometimes, the realization hits that the original goal of the meeting no longer makes sense. For example, maybe the discussion was built on a flawed assumption or misalignment. In that case, don't be afraid to abandon the original agenda and shift gears. Just make clear the new goal you will work toward. Letting the group flow toward what really needs to be addressed can be the most productive outcome.

Define and timebox the meeting structure

Set expectations up front on how much time you will spend on gathering ideas, discussing options, and making a decision.

Assign a timekeeper. It can be you as the facilitator or someone else in the group. The timekeeper's role is simple but important: keeping the pace and reminding the group when time is running out so you still have enough time to conclude properly.

That said, sometimes things don't go according to plan. If the decision needs more time, acknowledge it honestly. Summarize where you are, document the partial outcome, and propose a follow-up session.

Use visual tools to ensure alignment

One of my favorite techniques is whiteboarding. Use a whiteboard or a digital tool like FigJam or Miro so that everyone can contribute ideas, see what's being discussed, and track progress live. Visual tools help people process complex topics better and spot gaps or misalignments faster.

Another great tool is a RAID board, where you track risks, assumptions, issues, and dependencies.

Also, having a dedicated space to park side conversations or concerns (like a "Parking Lot" board) ensures people feel heard without derailing the main conversation.

Use different tools to capture ideas

Not everyone feels comfortable sharing ideas out loud. Some people might prefer writing them down.

So use multiple channels for input: anonymous forms, one-on-one discussions, or Post-it notes.

Separating the idea from the person makes it easier to have objective, less heated conversations around options and trade-offs. This creates a safer space for honest debate.

Constantly validate assumptions

Most misalignments happen because people think they are on the same page when they are not.

Make it a habit to regularly validate what you hear. Use simple phrases like "If I understand correctly, you mean…" to clarify.

It might feel repetitive, but catching misunderstandings early is always better than cleaning up miscommunications later.

Make agreements specific and document them

Before ending any meeting, take a moment to spell out the decisions made. I always say it out loud: "Ana will work on X. John will review Y."

I immediately write it down live in a shared document while sharing my screen with the group. This turns the meeting into a visible, living commitment. No more "I thought you meant…" conversations afterward. And funnily enough, very often someone will chime in at this point: "Wait, that's not what I understood." Exactly what you want to uncover before leaving the room.

Follow up

Without proper follow-up, even the best meetings turn into wasted time.

Always define the next step: maybe it's setting up a follow-up meeting; maybe it's agreeing when progress will be reviewed asynchronously.

Whatever it is, make it clear how and when the actions will be revisited. Otherwise, things get lost in the chaos of daily work.

Beyond structure and logistics, facilitation is also about navigating group dynamics, energy, and communication. These extra techniques will help you lead smoother, more inclusive, and more productive decision-making sessions:

Facilitation starts before (and continues after) the meeting

Sometimes, especially for complex topics, you'll need to help people show up ready. That might mean priming them with questions to think about in advance, reminding them to do the reading, or even summarizing key points in advance if you know your team doesn't always prepare.

If half the attendees show up unprepared, you'll have to decide whether to adapt in the moment, perhaps by spending time reviewing the materials together or rescheduling with clearer guidance.

Prioritize "moving forward together" over "being right"

Don't just focus on proving your solution is the best! In fact, if you can't get others on board, it might be a sign your solution needs rethinking.

What matters most is finding a path the whole team is willing to commit to. Alignment doesn't mean everyone agrees 100%, but it does mean there's shared ownership and a collective commitment to move forward together.

Consistency is often more valuable than the "correct" technical choice

If it's not an approach we absolutely want to move away from—because it's clearly outdated or fundamentally flawed—then sticking with what we used last time is often the better choice. This applies the most when we are talking about technical tools, although the rule applies more generally as well. Consistency reduces cognitive load, and having clear, consistent ways of doing things matters far more than chasing trends or perfect solutions.

It doesn't always have to be you facilitating

It actually makes your job way harder to be the tech lead, the facilitator, the developer, and the team member in the same meeting. Once things are running smoothly, I suggest training others in facilitation skills. It's a great growth opportunity for them and frees you up to contribute better during meetings.

With facilitation comes power, not just responsibility

People often forget that the role of facilitator doesn't just come with responsibility; it comes with power. The power to shape the flow of the conversation, to make space for quieter voices, to steer the group back when things go off track. It's subtle, but it's real.

And if you're stepping into the role of facilitator, especially as a tech lead, it's important that you use that power thoughtfully.

Use it to balance participation by asking people to raise their hands before speaking. This allows you to guide the conversation and ensure space for quieter voices, not just the most vocal.

Use it to create space for quieter voices, with prompts like "Would anyone like to offer a different perspective?" You can also gently invite someone directly—"July, what do you think?"—as long as you've checked beforehand that they're comfortable being called on.

If a conversation is getting too heated and going nowhere, don't be afraid to break the tension: "OK, pause! I see we all care deeply about this. How could we move forward from here?"

Use your facilitator role to keep momentum when a discussion starts drifting. If a new topic comes up that's not urgent but still valuable, you can say, "I see this is not something we can address right now, but it's a risk worth capturing. Maybe we can just add it on our RAID board for now and return to our main point?"

And when things go completely off course, don't hesitate to bring the group back with a simple "I feel we're derailing a bit here. Let's recap what we already discussed."

Manage team energy, not just time

In longer sessions or recurring meetings, your team's energy becomes just as important as the agenda. Fatigue leads to disengagement, rushed decisions, or conversations dominated by a few voices.

Be mindful of how people are doing. Plan for breaks. If you're in person, energizers and movement-based activities can help reset focus.

Even small gestures can set the tone. For example, I used to bring coffee and croissants for my team every time I got the chance during our sprint planning. It was an early morning meeting every two weeks, and this small routine helped start things on a positive, relaxed note.

The format matters too: switching from group discussion to silent writing or pairing exercises can refresh the team and maintain engagement.

The best ideas often come when people are relaxed and recharged.

Facilitating technical decisions well is one of the most useful and underrated skills of a tech lead. It's about creating the space where the best solutions can emerge and making sure your team is aligned enough to act on them.

TRACK TECHNICAL DECISIONS—ARCHITECTURE DECISION RECORDS

All the technical decisions you are making as a team should be tracked and made visible, to your current team and to any future team members who'll wonder "Why did we do it this way?" You want a record that captures how the decision was made, what trade-offs were considered, and what consequences were accepted.

This is where architecture decision records (ADRs) come in.

ADRs are a lightweight and effective way to capture the team's understanding at a specific point in time. They reflect the best decision made given the information available, along with the trade-offs considered. Instead of aiming for permanence, ADRs focus on documenting the context and reasoning behind a choice, so your future self, or others, don't have to guess why something was done.

Starting with ADRs

Start simple. The best way to begin using ADRs is by writing one for your next significant technical decision. Or even better, start by documenting a recent decision that's already been made. This shifts the focus to learning the process rather than trying to capture everything perfectly in the moment.

Don't overengineer it. Keep them lightweight and make it a habit. Encourage the team to use them anytime you're facing a decision that's architecturally significant, even if it feels small at the moment.

A simple ADR format might look like Table 9-3.

One way to make this process stick is to integrate ADRs into your workflow. For example, make writing an ADR part of the definition of done for certain stories. When a story includes a technical decision, require the team to capture it in an ADR before it's considered complete. This way, the habit becomes part of how you build.

Table 9-3. Simple ADR format

file name
ADR title
Give it a clear, descriptive name. Match the file name so it's easy to search.
Context
Describe what you knew at the time:
What constraints were you working under? or What challenges were you solving?
Stay neutral: just the facts.
List any alternatives that were explored and why they were rejected.
Decision
Clearly state what you decided. Use full sentences, in an active voice: "We will…"
Consequences
Spell out what this means, both good and bad. For example:
• This introduces tech debt we'll address post-launch.
• This approach supports X but makes Y more difficult.
• Although option Z was preferred by some, we agreed to go with this and revisit later.
Capturing consequences is often the most valuable part. It shows the trade-offs. It creates alignment. And it gives your future self a realistic picture of what you signed up for.
Extra information
e.g., Date, Status, Related ADRs, …

Use ADRs as a tool during conversations. Capture options discussed, trade-offs considered, and concerns raised while the decision is being made, not just after it's done.

Example of an ADR

You can see how a simple ADR might look in practice in Table 9-4. This example shows how to capture the context, the decision, and the consequences in a lightweight format that's easy to read, share, and update. The goal is to create a shared understanding your team can refer back to.

Table 9-4. Example of an ADR

adr-0001-use-postgresql-as-main-database.md
ADR-0001: Use PostgreSQL
Context
We need a reliable, open source relational database that works well with our tech stack (Node.js + TypeScript).
The product roadmap involves features that require strong consistency guarantees, transactional support, and a flexible schema.
Options considered:
PostgreSQL: Strong match for our needs. Already used in other services. Team has solid experience. Rich ecosystem and tooling.
MySQL: Similar to PostgreSQL but lacks some advanced features we may rely on (e.g., JSONB support).
MongoDB: Flexible and fast to prototype with but lacks the consistency guarantees and transactional support needed for billing-related features.
DynamoDB: Highly scalable but has a restrictive query model and would require a significant shift in tooling and team expertise.
Decision
We will use **PostgreSQL** as our primary database for the new billing service.
Consequences
✓ Alignment with team skill set and existing infrastructure.
✓ Strong ecosystem and tooling support.
✗ Higher operational complexity than DynamoDB.
✗ Less flexible for schema-less data, which we'll address with JSONB columns where needed.
✓ Easy integration with our current CI/CD pipeline and monitoring stack.
Note: Option Z (MongoDB) had some strong internal support, but we agreed to revisit if needs change.

Date: 2025-04-28
Status: Accepted
Supersedes: _None_
Related ADRs: _None_

ADRs challenges

There are several challenges you're likely to encounter when it comes to ADRs. Here they are:

Storing ADRs

Tech teams often debate whether to store ADRs in a single shared repo, spread them across individual services, or place them in tools like Notion or Confluence. The key is context: what problem are you solving?

A good starting point is to keep ADRs local to the service where the decision applies, in the same repository as the code. This makes them easy to find, keeps them versioned alongside the implementation, and allows your team to use the same tools for editing.

Over time, you might find it useful to move some ADRs to a centralized repo, especially if they apply to multiple teams, you're standardizing practices across the org, or tooling makes service-level storage less practical. This shift often happens when architectural guidance needs to be shared more broadly or when your company has specific guidelines for ADR storage.

Any approach works, as long as your team knows where to find them and follows a consistent process.

But what matters most isn't where you store ADRs; it's that you actually write them and make them accessible. Don't let the storage debate drain your team energy. Pick something and move forward. You can always move them later. At the end of the day, it's just text.

Updating ADRs

Keeping ADRs in the same repo as the code also makes it easy to update them. There are two common styles here: some teams update the original ADR file and use Git versioning to track changes. Others prefer to create a new ADR entirely and mark the old one as deprecated or superseded. Either approach is fine; the key is clarity. If you're writing a new ADR to replace an old one, make sure to add a note in the original: "This decision has been superseded by [ADR-0023-NewTitle]." That way, future readers won't be left wondering which guidance to follow.

When deciding which approach to take, consider your audience.

If your ADRs are read by people outside the engineering team, like product managers or business analysts, it may be better to create a new ADR and deprecate the old one. That way, the full reasoning behind each change is clearly visible and easy to understand without having to dig through Git history.

On the other hand, if your readers are comfortable with Git and prefer a compact history, updating the existing file might be the cleaner option.

Pick the approach that makes decisions easy to trace for your team.

What counts as "architecturally significant"?

Another challenge is knowing what's worth documenting.

Don't write an ADR for every small config tweak. Focus on architecturally significant decisions: things that shape how you work, structure your codebase, handle failure, or scale. Things like how you handle authentication, which database to use, changing from REST to GraphQL, how you approach retries and timeouts, or where you draw service boundaries.

If you're unsure, ask: "Would someone new on the team benefit from knowing why we chose this approach?"

Who owns ADRs?

Ideally, ownership is shared across the team. But, as always, when everyone is responsible, no one is, so it helps to clarify who's responsible for maintaining ADRs, whether it's the person driving the decision or a rotating role. Don't let ADRs become stale just because no one feels accountable.

Besides documenting decisions, ADRs also act as a kind of contract, a shared commitment from the team to follow a chosen path, knowing the trade-offs. They help you move forward even when there's disagreement, by acknowledging concerns and leaving space to revisit them later.

For example, imagine two developers arguing over which JSON library to use. If the team already made a decision months ago and documented it in an ADR, with the options considered, trade-offs, and rationale, it can help settle the discussion quickly. Of course, if new context emerges, you can revisit the decision. But in many cases, it saves time and prevents rehashing debates over choices that were already made thoughtfully.

Used well, ADRs are a low-effort, high-impact way to preserve team knowledge, align on decisions, and reduce friction as your product and team evolve. Especially in fast-moving environments, they're one of the simplest ways to protect your team's context, and one of the easiest habits to adopt that pays off in the long run.

Encouraging Technical Excellence

Technical excellence doesn't just happen. It's the result of shared, intentional decisions, how we build, how we review, how we test, and how we operate. As a tech lead, your role is to create the environment where those decisions are made deliberately and consistently.

Your job is to keep the bar high by making it visible, agreed upon, and reflected in how your team actually works. That means setting clear expectations, building strong engineering habits, and reinforcing the idea that quality is a shared responsibility.

In this section, you'll learn how to define standards that feel shared and meaningful, ones your team actively commits to, instead of rules handed down from above.

I'll also cover how to build fast feedback loops through integration and deployment, how to use measurement to guide thoughtful improvement, and how to make testing a consistent and valuable part of everyday development work.

DEFINE TEAM STANDARDS

As a tech lead, you play a key role in helping your team align on what "good" looks like, technically and operationally, as well as behaviorally.

You may already have company-wide guidelines or platform-level constraints, but within your team, you still need to define your own local standards. These are the agreements that help your team move fast together, without reinventing the wheel for every decision.

When we say "standards," we don't just mean linting rules or test coverage thresholds. We're talking about agreements at every level of how your team operates:

- How decisions are made (see the section "Help Your Team Make Technical Decisions" on page 272)

- How the system evolves (see the section "Build a Technical Vision (and Turn It into a Strategy)" on page 269 and the section "Track Technical Decisions—Architecture Decision Records" on page 281)

- How code gets written, reviewed, and shipped (see the section "Aim for Fast Feedback" on page 287 and the section "Ensure Quality Through Testing" on page 290)

- How priorities are chosen and delivery is tracked (see the section "Planning and Keeping a Project on Track" on page 291 and the section "Balancing Multiple Streams of Work" on page 297)
- How you work together as a team (see the section "How to Build a Healthy Team Culture" on page 185)

These agreements form the invisible scaffolding of your day-to-day work. When they're unclear or inconsistent, the team slows down. People second-guess what "done" means. Decisions take longer. Reviews are frustrating. Quality drops.

But when your standards are clear and shared, your team can work independently without diverging. You don't have to be in every conversation or review every line of code. Because the expectations are understood, decisions can scale.

As a tech lead, don't define these standards alone; create space for the team to define them together. That might mean running a workshop, reflecting in a retrospective, or revisiting old documentation that no longer reflects reality. It's less about writing rules and more about surfacing assumptions and creating alignment.

One trap to watch out for: having overly aspirational standards that no one actually follows. A great rule of thumb is this: your standards should reflect what you're willing to review and uphold, not just what you hope for. They should be visible, actionable, and revisited regularly, especially as the team changes or the product evolves.

You don't need to solve every disagreement up front. But do agree on a mechanism for resolving them. Standards evolve, and that's healthy. What matters is that your team knows where to look, how to propose a change, and how to hold each other accountable.

AIM FOR FAST FEEDBACK

In Chapter 8, I talked about the value of having great processes for continuously delivering high-quality value to your customers, and shipping code that's been tested, validated, and is up to standard.

But even with the best processes, things will go wrong. That's inevitable. And the only way to know *when* they go wrong, and *how fast* you can react, is by having the right observability in place. Metrics, alerts, and alarms are your early warning system. Without them, you're flying blind and will discover issues only once they've already impacted users.

As a tech lead, you're not expected to be an observability expert, but you are expected to care—making sure the right signals are in place and that people are paying attention to them. You're responsible for asking the right questions: Do we actually know when something's broken? Are we measuring what matters? Can the team respond quickly when things go wrong?

Your role is to guide the conversation. Help your team align on what "good enough" observability looks like. Push for clarity over coverage. Facilitate decisions about what to monitor, what can be dropped, and how to stay on top of what matters as your system evolves. Think of yourself as the person connecting the dots: from business impact to system behavior, from vague symptoms to root causes, and from noisy signals to actionable alerts.

The first issue I've seen with observability is the most obvious one: no data. Just not enough visibility into what's happening in your system to make any useful decisions.

But more often, the bigger problem isn't too little data; it's too much. Any observability tool you hook up to your services will come with dozens of automatic metrics. Suddenly you've got too many dashboards, too many graphs, and too many alerts, many of them noisy, low-value, and constantly blinking yellow. You stop trusting them. You ignore them. And when something really does go wrong, it's hard to see the signal through all the noise.

This happens because teams treat observability like a side task, something separate from product work. Someone adds a few metrics "just in case." Another person adds some alerts without checking if they're actually useful. Rarely do we go back and clean things up. Rarely do we ask: is this data still helping us?

As a tech lead responsible for a product, I basically want to know two things: is it available? and does it work as expected?

I remember a time when our dashboards were filled with charts no one could explain. We had alerts firing at random times, most of which we ignored. Then one weekend, something actually broke, and it took us hours to realize it because the real issue was buried under noise. That was the moment I realized we needed fewer, more meaningful signals, metrics that actually meant something to us.

If you're dealing with observability noise, here's my suggestion: start fresh. Forget your current dashboards for a second. Sit down with your team and ask, "What do we actually need to monitor? What's a real problem? What are the alarms we would wake up in the middle of the night for?"

When your team defines what matters together, you create clarity around what "good" looks like, what constitutes a real issue, and where to focus during incidents. This shared understanding can dramatically reduce noise, speed up incident response, and build a stronger sense of ownership across the team.

Start from there. A good rule of thumb: one dashboard per product, not per service. All the key metrics should fit on a single screen. If you glance at it and can't immediately tell whether things are OK, something's wrong. And the metrics need to be meaningful. Metrics aren't useful unless they tell you when something is wrong.

This requires a mindset shift. Instead of asking "What should we measure?" ask "When is it actually a problem?" Define your service-level indicators and service-level objectives. Set limits. Tune/refine/iterate on thresholds. And be courageous enough to delete what doesn't help. If you have an alert that's always red but you never act on it, do you really care about it? Probably not.

Modern observability tools like Datadog and Grafana can often show you historical data, even for metrics you didn't pre-configure, so don't be afraid to remove noise. If you need something, you can add it back.

Another strong habit: make observability part of your daily development work. Every feature, every task, should come with a conversation: "Do we need to monitor this? What does success or failure look like?" Add that to your definition of done.

Good observability helps your team catch problems early and turn those problems into learning opportunities. When something breaks, a system with the right metrics and logs gives you clues about what happened, what went wrong, and why. This makes incident reviews more useful and less about finger-pointing. Over time, this turns incidents into learning opportunities. You start to see patterns, connect the dots, and use that insight to prevent future issues. That's when observability becomes more than a safety net.

Observability is how your code talks back to you in production. If you can't hear it clearly, you're flying blind. Your goal isn't more data; it's faster feedback and better decisions.

The tech lead plays a critical role in shaping observability culture. Your job is to make sure the right metrics and dashboards are in place, that they reflect what truly matters, and that the team is using them to inform decisions. This means encouraging regular conversations about what signals are meaningful, keeping dashboards focused and actionable, and helping the team build habits around learning from data, not just collecting it.

ENSURE QUALITY THROUGH TESTING

Managing technical projects means driving quality throughout the entire development process. Testing plays a key role, preventing bugs and building confidence in the code, the process, and the team.

I already explored in detail how to build and evolve a healthy testing strategy in the section "Continuously Testing" on page 245. If you haven't read that section yet, that's where you'll find actionable guidance on defining testing standards, addressing common bottlenecks like flaky tests and slow pipelines, and building alignment with QA teams.

Here, the reminder is this: testing is a continuous investment in quality. Make sure your tests are useful, provide meaningful feedback, are trusted by the team, and support the pace you need to deliver with confidence.

Testing is something you continuously lead. As a tech lead, keep asking yourself: can we release with confidence? If something breaks, will we know? And can we fix it quickly, without blame? The answers to these questions are signals of whether your team has a reliable, quality-focused delivery culture.

If the answer to any of these is no, that signals a broader quality and leadership challenge. You have to make sure testing is a core part of the way work gets done, not something tacked on later. If tests exist but aren't helpful, or worse, slow things down, question them. If flaky tests are draining trust or key areas of your system lack test coverage, step in. And always make sure your testing approach matches the level of risk in the business and the speed your team is expected to deliver at.

Testing exists to support delivery. If it's creating friction instead, that's a signal to pause, reflect, and adjust it with your team.

Let the detailed strategies in the section "Continuously Testing" on page 245 guide your implementation. Use this section as a reminder that how well your team tests is a reflection of how well you lead quality.

Common Project Management Challenges

So far in this chapter, I've focused on aligning your team around shared strategies, building a strong technical vision, and encouraging technical excellence through quality, feedback loops, and decision-making practices. But even with all of that in place, projects still go off track. That's where your role as a tech lead becomes even more critical.

In this section, I'll shift focus to the everyday realities of delivery: how to plan and keep a project on track, how to react when a project starts falling behind, how to juggle multiple streams of work without overwhelming your team, and how to identify and mitigate the technical risks that could quietly derail your delivery.

Strong technical leadership means leading with clarity, navigating complexity, and keeping delivery moving, without burning out your team. You don't need to become a project manager to do that, but you do need the tools and mindset to guide your team through uncertainty.

PLANNING AND KEEPING A PROJECT ON TRACK

Most projects derail because of bad planning. And bad planning usually comes from a lack of proper communication at the right time between all the parties involved or from a lack of agreement.

I truly believe the best planning happens when everyone is involved from the beginning: product, business, stakeholders, developers, UX, QA, infrastructure—anyone who touches the product. Maybe not everyone is equally involved at every step, but they should all be present. I'm a big advocate for the whole team being part of this process, not just the tech lead. I've seen how much better the outcome is, how much more engaged the team feels, and how much better the product becomes. I've seen entire product strategies shift for the better simply because a developer asked a question early on.

But I also know that's not always the case. In many teams, not even the tech leads are brought into planning. A lot of tech leads still think planning is something product or project managers do and that business will eventually just hand over a list of things to build. That's how many teams still operate. Unfortunately. And when that happens, it often leads to plans that ignore technical complexity, result in unscalable or overly complex designs, miss important edge cases, or underestimate effort entirely. Without the right voices in the room early, you're more likely to build the wrong thing, slowly and painfully.

If you're lucky enough to be involved early, great. Take that opportunity. If you're not, ask why. Often the answer is something like "Oh, I didn't know you wanted to be involved." And just like that, you're in.

If you want to maximize the chances of success for any plan your team is leading or involved in, there are a few key things you need to make sure are in place:

Clear problem to solve/purpose

The best planning starts from a shared understanding of context and purpose:

- Why are we doing this? What's the problem or opportunity we're addressing?

- Who are the stakeholders? What do they care about? Who will use the solution, approve it, or be affected by it?

- What's the vision? What would success look like, not just in features delivered but in impact created?

From there, the team can envision the shape of the solution, whether that's architectural direction, key design principles, or a prototype, and only then move toward implementation planning.

Clear owner

I once helped a startup troubleshoot why a company-level project had missed its deadline by three months. The project involved three different engineering teams. My first question was simple: who owned this project? The responses revealed the core issue. The tech lead assumed the product manager was in charge. The product manager thought it was the engineering manager's responsibility. The EM pointed to the CTO. And the team believed the tech lead was managing it. Everyone assumed someone else was driving, but no one ever said it out loud.

The result? No one had a full picture of what was happening. There was no shared timeline, no one enforcing the deadline, and each team was planning and prioritizing independently on different boards. Nobody was coordinating the bigger picture.

The lesson was clear: when everyone is responsible, no one is.

One of the most common reasons I've seen tech projects and initiatives fail is the absence of clear ownership. You need one single point of contact: someone responsible for tracking progress, updating communications, following up on decisions, and reacting when things go off track.

Owners aren't expected to do everything themselves or be present in every meeting. Their role is to provide visibility into the plan, tracking progress, highlighting risks, and making sure responsibilities are clearly understood. With the right structure in place, the team should be able to move forward independently, without needing constant supervision.

Clear steps

Break down the work into tangible, achievable chunks. Each step should be actionable and easy to understand. Avoid vague labels like "implement feature." Instead, be specific: "build endpoint for X" or "create UI component for Y." This helps the team track progress and makes handoffs smoother.

At the high level, start with stakeholder conversations and product goals, then shape these into large epics that reflect business priorities.

From there, work with your team to split those epics into smaller tasks that can be executed within your sprint cycles.

Keeping this dual-layer structure, strategic (epic-level) and tactical (task-level), helps ensure everyone from business to engineering stays aligned on both the big picture and the next step.

Clear milestones

Define what success looks like at various points in the project. These are your checkpoints, moments where you pause, reflect, and potentially readjust. Milestones might signal the end of a sprint, the completion of a critical feature, or the delivery of a first internal demo. They help track progress, but more importantly, they create space to check alignment, surface risks, and make informed adjustments. Choose milestones that show meaningful progress, especially to stakeholders, and use them to validate or update your delivery plans.

Clear timelines

I've seen tech leads hesitate to enforce timelines, worried they'll come across as pushy or create pressure on the team. So instead, they say nothing. They go with the flow, hoping things will somehow be delivered on time.

But the truth is there is always a timeline.

The business side always has expectations about when they want something delivered. If it's not clearly stated, ask for it, and don't give up until you get a real answer.

If you struggle to get one, propose your own. Start by giving an intentionally long estimate, something you know won't be acceptable, like "OK, so we'll aim to get this done in about six months." Chances are, you'll get a reaction: "Six months?! That's way too long." Perfect. Now you can ask: "What would be acceptable?" Maybe they say, "We were hoping for three." Great. Now you have a target and something to plan around.

If it's not clearly stated, it becomes invisible. And when it's invisible, people assume they have all the time in the world. That's when you end up with someone refactoring a huge piece of the product in what was meant to be a small task. Without clear boundaries, people set their own. And everyone's idea of a reasonable timeline is different.

I remember working with a tech lead who was anxious about an upcoming delivery. Her team was behind schedule. Stakeholders were checking in constantly. She felt the pressure mounting but didn't want to push the team. "What's the actual deadline?" I asked. "There isn't one," she said. "But everyone expects it by May 1st."

"So there is a deadline," I said. "It just hasn't been made explicit." I suggested she share that with her team, not as a demand but as transparency. Just share that there's an expectation, explain why it matters, and bring the team into the strategy conversation.

She did. In our next one-on-one, she said, "That worked like magic. Once I explained the situation, the team prioritized the work together. We're actually ahead of schedule now."

This is the power of clear expectations. When people understand the "why," they align fast.

The key is to treat deadlines and milestones as shared goals, not demands. "We want to finish this by May 1st because of X." Then open the conversation. Is that realistic? Can we shift it? What's the impact if we don't? On us, on the customer, on the business? If the timeline is fixed, can we adjust the scope? What's the smallest version we could ship that still brings value? Could part of it be manual for now?

Creating this plan doesn't fall entirely on your shoulders, and usually you are not the one that drives it. In most teams, it's a shared effort between you and your product owner, product manager, project manager, engineering manager, or whoever else is involved. You all bring something different to the table. Your role is to contribute the technical perspective: how feasible things are, where the risks lie, how the team works in practice. Their role is to bring business context, customer needs, and priorities. Together, you're jointly responsible for making sure the team has a clear, realistic path forward.

A clear plan also enables speed. Once your team moves into execution, they should know exactly what they're working on and why. Nothing hurts productivity more than lack of focus.

I once joined a team full of motivated, capable engineers who were struggling to deliver, not because they lacked skills but because they lacked clarity. The backlog was messy, stories were vague, and tasks were made up as they went. Junior developers were blocked, waiting for direction.

When I joined as a tech lead, I quickly realized this wasn't a problem of motivation or talent; it was a planning problem. There was a product manager, but she was stretched thin, managing multiple teams and projects, and didn't have the capacity to go deep into the day-to-day planning. So instead of jumping straight into the code, I dove into the backlog. I started working closely with the product manager, legal department, and other teams involved, helping shape a clear high-level strategy. I set up planning ceremonies like pre-planning and refinement sessions and brought the team into the process.

In just a month, everything changed. The team was thriving. Work was moving fast, confidently. No second-guessing, no confusion, just clear priorities, a shared process, and a rhythm for getting things done. Once all of that was in place, once we had a clear plan, a clear timeline, and full alignment, I finally went hands-on to help deliver. Because at that point, the team had the structure and clarity they needed to actually move fast. Planning is what made that possible.

Another thing that kills focus is trying to do too much at once. If everything is a priority, nothing gets done. A common trap for teams is having too many work items in progress at the same time, resulting in jumping between tasks, starting but not finishing, and losing track of priorities. That's why part of effective planning is limiting WIP (work in progress).

Help your team stay focused by agreeing on how much work can be in motion at once. Encourage finishing over starting. Before picking up a new task from the to-do list, pause and talk about trade-offs, especially if there are still plenty of tasks in progress. In one of my teams, we had this principle so deeply embedded that if the board was full of in-progress tasks, no one would start something new. Instead, anyone without a task would pair with someone else to help finish what was already underway.

To wrap up, treat planning as a tool for collaboration. Great planning builds shared understanding, aligns expectations, and creates space for ongoing recalibration. That means having regular conversations at both strategic and tactical levels, with stakeholders, your team, and anyone whose work intersects with yours. When you lead planning this way, you'll catch risks earlier, adjust faster, and build better things together.

Plans should be visible, dynamic, and shared.

RESPONDING TO PROJECT DELAYS

Even with the best planning, things will go wrong. An incident might hit. Someone gets sick. You run into a technical complexity you underestimated. Or a third-party dependency slips.

That's normal. What matters is how you respond.

First thing to do as a tech lead: make the risk visible. The moment you suspect a delay or see something slipping, your first responsibility is to make the risk apparent. Don't wait. Share it with your team, your stakeholders, and anyone else who might be affected. Then bring everyone together to revisit the plan. Can you shift the timeline? Trim scope? Deliver part of the work manually for now?

In one project, we were launching a new product that had already been sold to customers with a promised start date. Two weeks before launch, we realized the third-party API we were relying on wouldn't be ready in time. Postponing wasn't an option, so we got creative with scope.

We identified the minimum set of features that absolutely needed to work automatically.

Then, we renegotiated with the third-party provider to deliver a smaller, stripped-down version of the API that could support just those essentials.

For one feature, we made a conscious decision to handle it manually at first. It wasn't scalable long term, but with only a handful of early clients and the feature being triggered manually, it was totally manageable. We added a flag to indicate when the manual process was needed, and aligned with the PM on a follow-up plan to automate it down the line.

The launch went smoothly thanks to our ability to adapt quickly, communicate clearly, and make thoughtful trade-offs, even under pressure.

You're not responsible for solving every problem alone. Your focus is on maintaining alignment across people and priorities: keeping communication open, translating business goals into delivery plans, and ensuring the team understands what's expected before it becomes a surprise.

Also, be careful with the idea of "just adding more people." When a deadline is at risk, stakeholders often say, "Can't we just add more developers?" On the surface, it sounds like a quick fix. It sounds logical: more hands, faster delivery. But in reality, onboarding new developers slows things down. Not only do they need to learn the codebase; they need to learn how your team works, how you collaborate, how you deploy. It's like the old saying: what one developer can do in one day, two can do in two days. Unless you're already set up to scale with new people, it's rarely a quick fix.

Second thing to do as a tech lead: protect the team's focus. As a tech lead, part of your job is to shield the team from distractions and last-minute "urgent" requests that pop up mid-sprint. You won't be able to block every interruption, but you can create clarity around what the team is working on and why.

Without this kind of buffer, teams can fall into what some call the "interrupt-driven death spiral," constantly reacting and context switching and never finishing meaningful work. Make sure that doesn't happen.

Make your priorities visible. Encourage your team to talk openly about priorities when unexpected tasks get thrown their way, and to explore options like saying "Not right now, because we're focused on delivering X" or "Let's put it in the backlog and we'll revisit later," rather than jumping in automatically.

If you don't set these boundaries, everything becomes a priority, and then nothing gets done on time. Being the buffer between the team and the chaos is one of the most valuable ways you support delivery.

Another thing to keep in mind during these pressure situations: checking in on morale. When a project's behind, stress rises. Tensions run high. In moments like this, one-on-ones are your best friend. Use them to check how people are feeling. Are they worried? Burned out? Feeling stuck or unsupported? This is where you catch problems before they explode. These conversations help you know when to push, when to shield, and when to pause and reassess.

In short, when a project slips, don't panic. Slow down. Communicate. Recalibrate. And support your team. That's how you keep moving forward without losing trust or burning out your people.

BALANCING MULTIPLE STREAMS OF WORK

If your team owns a single product with clear boundaries, the same tech stack, familiar tooling, and consistent services, you're in a fortunate position. Many tech teams today aren't that lucky. It's increasingly common for tech teams to be responsible for multiple products or projects spread across different systems, services, programming languages, and infrastructure. And this comes with a cost.

Everything takes longer than planned: onboarding, maintenance, debugging. The biggest challenge is context switching. Jumping between tech stacks or products drains focus and energy. It slows the team down and makes execution more difficult.

I've been there. My team was responsible for two completely different products, both high-priority, both business-critical. But one lived in a legacy .NET monolith running in a datacenter with RabbitMQ and MySQL. The other was a modern Scala-based microservice using Kafka streams and DynamoDB and deployed in AWS. The mental overhead of switching between these two worlds was enormous.

And yet, we made it work, by factoring this challenge into how we worked as a team.

We used tools to reduce context-switching friction: things like clear onboarding guides, well-maintained internal wikis, and architecture diagrams that gave fast overviews of the systems involved. This was not a setup where relying on code for documentation alone was an option.

Our delivery process was smooth, built on continuous deployment and supported by thorough testing and strict quality standards. We explicitly accounted for context switching in our estimates, acknowledging that jumping between products carried a cost.

We also leaned heavily on pair programming. Any time someone had to switch to the other product, they weren't doing it alone; there was always someone else to carry the context, ease the transition, and help get up to speed. Constant, structured communication helped keep us aligned, internally and with stakeholders. We maintained a clear sense of strategy and focus.

I also was regularly checking in with the team to make sure this approach was still working. Was the context switching becoming too much? Were people feeling overwhelmed? These conversations helped us stay ahead of problems before they grew. I knew that what worked at one point might stop working as priorities shifted.

Even with all this, it was hard. We struggled from time to time. But we approached it as a learning opportunity. Where else could we have explored two completely different technology stacks, back-to-back, as part of the same team? It stretched us, but it also grew us.

There are a few other approaches I've seen work well for other teams in similar situations.

Clarify ownership boundaries up front. One of the most helpful things you can do is explicitly define who owns what. Which team is responsible for which product or service? What does that ownership include—feature work, support, maintenance? Where do responsibilities stop? This removes ambiguity and helps avoid surprises when work overlaps.

Instead of juggling both products at the same time, you can place one product into full maintenance mode temporarily, agreeing to handle only critical bugs or issues, while prioritizing the other for a defined period. This would allow deeper focus without letting anything fall through the cracks.

Proactively challenge the assumption that your team has to keep owning everything. If the load is too high, it's OK to hand something off. A clear, structured handover to another team, or even a temporary reallocation of ownership, can create space to breathe. In our case, we had this conversation and decided we wanted to keep both.

I've also heard of teams using rotating sub-teams or specializations to manage multiple projects, though I haven't tried it myself. The idea is to split the team temporarily—one group focuses on Product A, another on Product B—then rotate members regularly. It's supposed to reduce context switching while spreading knowledge gradually. Seems like a smart way to balance focus and flexibility.

And of course, escalate early. If your team is overwhelmed, don't keep it to yourself. Talk to your stakeholders. Show them the impact of divided attention: slower delivery, more bugs, rising frustration. Bring the problem to light while there's still time to do something about it.

Balancing multiple streams of work will never be easy. But with clear priorities, deliberate trade-offs, and a willingness to set boundaries, it becomes manageable. If you're in this situation now, take a moment to assess, then try one of these adjustments. It might be just what your team needs.

MANAGING TECHNICAL RISK

Every project comes with risks. Your role as a tech lead is to ensure these risks don't transform into issues.

A risk is something that *might* happen. It's a potential issue. For example: "A server crash during peak hours might lead to data loss."

An issue, on the other hand, is something that has already happened, like a problem in production. For example: "During peak hours, the search page takes four times longer to load."

Issues are dealt with reactively: you fix them, work around them, or accept them. For instance, you might increase the number of search servers temporarily to reduce load time. For a longer-term solution, you could restructure the search index to better match common user queries. In some cases, the best choice might be to simply acknowledge the issue, such as showing users a message when search is slow, especially if the impact is minor and the cost to fix is too high right now.

Risks, however, must be identified, tracked, communicated, and planned for.

Identifying risks

Every project comes with risks, and no, you won't be able to mitigate them all. But you can make them visible and prepare for the most critical ones using different techniques.

One useful technique is Agile threat modeling, which focuses on identifying security risks. Some companies have a dedicated security team, but they're often stretched thin. That means security reviews come too late, miss key context, or rely on noisy automated tools.

Agile threat modeling flips that: it's done more frequently, in smaller chunks, and involves the whole team. It becomes part of your backlog and planning. I first encountered this practice at Thoughtworks. They've even published a full guide on how to run one: the Agile Threat Modelling Workshop Guide (*https://oreil.ly/jDXcb*).

Another excellent tool is risk-storming. It's a collaborative, visual exercise where the whole team, devs, project managers, product managers, designers, and stakeholders identify risks together. You map them out, prioritize them, and discuss mitigation strategies. The guide at *riskstorming.com* is a great place to start.

And if you've been maintaining a RAID board (risks, assumptions, issues, dependencies) throughout your decision-making or solution brainstorming sessions, this is a great moment to bring it up again. It can provide a clear view of previously identified risks and ensure continuity in addressing them proactively.

Identifying risks should be a regular part of your project rhythm. A good rule of thumb is to consider risks at the following times:

- At the start of a project, to uncover early architectural or security concerns

- Before major milestones or deliveries, to prepare for potential blockers

- After significant changes in direction or scope, when new risks often appear

- On a regular cadence, such as monthly or at the start of each sprint cycle, to keep the conversation alive

You don't need to run a full workshop every time. Sometimes a quick review of your existing risks and assumptions is enough.

Mitigating risks

Once you've identified your risks, you need to evaluate them. A common way to do this is using a 2 × 2 matrix:

Probability
> How likely is this to happen?

Impact
> If it does happen, how bad is it?

Based on those two factors, you can choose a mitigation strategy (Figure 9-1).

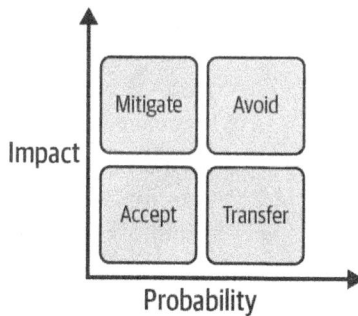

Figure 9-1. Risk-mitigation strategies

You can choose to mitigate, avoid, accept, or transfer the risk:

Mitigate
> Take steps to reduce the likelihood or impact.
>
> Example: You rely on a third-party payment provider, and downtime on their end would block customers from completing purchases. You mitigate this risk by building a lightweight fallback system: when the provider is down, customers can still place orders, and payments are processed once the service is restored.
>
> You can't eliminate the risk of the third-party service failing, but you reduce its impact on the business. Instead of losing revenue or frustrating users during downtime, you create a way to keep things moving while minimizing disruption.

Avoid

Change your approach so the risk no longer applies.

Example: You were planning to allow users to upload and preview arbitrary file types in your application. But this opens up a range of risks, like malicious file uploads, virus-laden attachments, or vulnerabilities in previewing logic.

After assessing the effort required to scan and sanitize all file types securely, you change the approach: users can upload only PDFs, which are stored without preview and scanned with a standard antivirus tool before storage.

Allowing arbitrary file uploads introduces complex security concerns that are difficult to fully mitigate. By narrowing the scope to a safer file type and removing the preview feature, you avoid exposing your system to these risks altogether.

Accept

You're aware of the risk but decide not to act, at least for now.

Example: You know your CI pipeline is slower than ideal, but improving it would require infrastructure changes that aren't a priority right now.

You're aware this might lead to longer feedback loops for developers and reduced iteration speed. It could eventually affect team morale or delivery timelines. But for now, the impact is manageable, so you choose to accept the risk, monitor it, and set a reminder to visit it in some time.

Transfer

Shift responsibility elsewhere, like outsourcing or involving another team.

Example: You're worried about the security risks in handling user authentication in-house. Instead of building and maintaining your own auth system, you decide to use a third-party identity provider.

You're aware that authentication is a critical and sensitive area; getting it wrong could lead to security breaches and compliance issues. But instead of carrying that risk internally, you transfer it to a provider with better expertise, support, and security practices in this area.

Choosing how to respond to risk is just as important as identifying it. Whether you accept, transfer, mitigate, or avoid a risk, the key is to do it intentionally, with a clear understanding of the trade-offs involved.

Not every risk needs action right away, but every risk needs visibility and ownership.

Tracking risks

After a risk-storming or threat modeling session, you'll have a list of prioritized risks. But that's just the beginning. You need to track them.

A risk log can be a Google Doc, a Jira board, or a spreadsheet; it doesn't matter, as long as it's reviewed regularly. Each risk is clearly described and easy to refer to in conversation. The mitigation plan and current status are kept visible so that anyone on the team can quickly understand where things stand. And each risk has a clear owner—not necessarily the person executing the work but someone responsible for making sure it's tracked and acted on.

For risks you've decided to accept, set a reminder to revisit them. Add a review date and use a system that will surface it back into your workflow.

Don't let this list get out of hand. Focus on the top-impact, high-likelihood risks. Keep it lean. A living document is valuable only if it stays useful.

Communicating risks

The value of tracking risks comes from the conversations and actions that follow.

The tricky thing is that not everyone sees risks the same way. A product owner might not care about a rare JSON parser vulnerability, but your security lead absolutely will.

That's why communication matters. When talking about risks, do it early, especially with supporters and decision makers. Don't just surface the problem; come with a clear plan and a few alternatives. Focus on the impact: explain what could happen and why it matters. Be empathetic. Try to see things from their perspective: what would they worry about? And be honest. Don't exaggerate to make a point, but also don't downplay something important. Focus on facts.

Managing risk is as much about alignment as it is about action. Use the tools and conversations described here to stay ahead of problems and to help people see clearly what's coming.

Key Takeaway

Managing technical projects is about more than tracking tickets or hitting deadlines. It's about creating clarity, reducing ambiguity, and helping your team move forward with focus.

From aligning on a shared strategy and defining a clear technical vision to making decisions efficiently and documenting them well, strong technical leadership gives your team the structure it needs to deliver.

You won't be able to plan for everything. But with the right habits—clear planning, fast feedback loops, ongoing measurement, and continuous communication—you'll be able to keep projects on track, navigate complexity, and recover quickly when things go off course.

10

Bringing It All Together: Navigating Technical Leadership

Technical leadership is a balancing act. It's a role that demands clarity, empathy, technical depth, and strategic awareness, often all at once. Throughout this chapter, I'll look at how to bring these elements together in practice as interconnected parts of your day-to-day work.

I'll start by exploring one of your most high-leverage responsibilities: growing the people around you. Next, I'll shift focus to stakeholder management. Your ability to represent your team, align across departments, and communicate technical ideas clearly to nontechnical partners is a defining part of the role. I'll walk through how to identify the right stakeholders, manage those relationships intentionally, and adapt your communication to influence effectively.

Finally, I'll unpack some of the most common, and often unspoken, challenges in aligning tech with the business. Whether it's getting buy-in for technical decisions, figuring out how your impact is measured, or managing the emotional and mental load of the role, this chapter offers practical strategies for navigating the messy, high-stakes reality of tech leadership.

This is where everything comes together: your technical voice, your people instincts, your strategic thinking, and your ability to sustain it all without burning out.

Developing and Growing Technical Talent

Growing the people around you is an expectation that usually comes with a senior role. As a tech lead, it's about taking this to the next level.

This doesn't necessarily mean doing more pairing or supporting more individuals directly. It's about stepping back and thinking more strategically about your team's growth: leveraging team-wide opportunities for development rather than focusing only on individuals. It also involves greater influence over things like budgets and the types of learning support your team can access.

With that broader perspective in mind, this section will explore how to identify your team's learning needs, support their technical growth with the right tools and opportunities, and help them strengthen the soft skills that turn capable engineers into trusted collaborators and future leaders.

ASSESS AND IDENTIFY LEARNING NEEDS

As a tech lead, part of your responsibility is to make sure your team has the right mix of technical and soft skills to deliver on what's expected. That starts with being intentional about understanding where your team is today and where they need to grow.

Start by looking at your context. Are you in the middle of a migration? Building something new? Maintaining legacy systems? Each of these requires different strengths. Your team might be great at launching new projects but struggle with long-term maintenance. Or the other way around.

I once led a team that got the opportunity to build a brand-new product. It was a high-visibility project, and we were excited to take it on. But the catch was this: it was a prototype. We had to move fast, prove the product's viability, and skip over things like full automation and perfect test coverage.

Our team came from a different world. We were used to working on a revenue-generating monolith with a "do it right" mindset: TDD (test-driven development), deep up-front clarification, robust processes. This new initiative demanded speed, iteration, and scrappy fixes, completely the opposite of what we were used to.

We took on the challenge, but the friction was real. Team members were uncomfortable with the shortcuts. Stakeholders were frustrated by delays. Everyone was out of their comfort zone. After some settling-in time, we started to find our rhythm. But it took deliberate effort, including lots of conversations and iteration. This taught me a valuable lesson: before you take on work that demands a different mindset and skills, assess whether your team is equipped for it, and if not, plan for how you'll support them through that shift.

That's why it's worth doing regular checkpoints with your team, especially before jumping into a new challenge. Here are some questions you can explore together to assess where you are and where you want to grow next:

- What parts of our system feel fragile or hard to change?
- Are there areas where we rely heavily on one person's knowledge?
- What's something new we've wanted to try but haven't felt ready for?
- What patterns do we notice in our recent incidents or blockers?
- Where do we feel like we're guessing rather than deciding with confidence?
- What are we curious to learn more about?
- What skills would make our day-to-day work smoother?
- How do we currently share knowledge, and what's missing?
- What would help us grow more effectively as a team?

Once you've gathered this input, look for patterns. Are there recurring challenges? Common frustrations?

All of these give you clues about where the learning gaps are.

Don't forget to balance team-wide patterns with individual needs. Every developer is on a different growth path. Some may be ready to step into mentoring roles or lead initiatives. Others may need support just getting comfortable with basic technical tasks. Hold both levels in mind: what the team needs as a whole and what each person needs to thrive.

Next, align those learning needs with what's coming. What features are on the roadmap? Is there a big migration planned? Are you moving to the cloud or adopting a new architecture? You want to invest in skills that help the team deliver value, not just what's technically interesting. For example, if the business is focusing on an upcoming cloud migration, it might not make sense to invest time in exploring a new web framework right now.

Once you've identified the most important learning areas, make a plan. You don't have to fix everything at once. Pick one or two priorities, set a goal, and define how you'll track progress. For example, if you want to build a stronger feedback culture, measure how often feedback conversations are happening. Make it visible, and revisit it often.

Continuously upskilling your team is a team effort. Involve your team in assessing needs and shaping priorities. Partner with your product manager or engineering manager to align growth with delivery needs. Use one-on-ones to validate direction, get personal insight, and ensure that growth goals reflect both individual aspirations and the evolving needs of the business.

By consistently assessing and addressing your team's learning needs, you increase the team's ability to deliver and adapt to any context.

TOOLS FOR UPSKILLING THE WHOLE TEAM

Once you've identified the growth areas for your team, it's time to act. There are many ways to upskill your team, but not every method fits every context. Your job as a tech lead is to be intentional about which tools to use, when to use them, and how to integrate them into the team's daily work.

If your entire team needs to build technical strength in a particular area—say, a new framework, architectural pattern, or development practice—focus on collective learning. Here are a few tools you can use:

Dedicated learning time

Talk to your manager or stakeholders to secure protected time for learning. This could be a learning day each sprint or a short-term drop in delivery expectations.

One simple strategy is to integrate learning time directly into your estimations. Just like you'd account for code reviews or testing, learning has an impact on delivery. Making it explicit in planning helps set the right expectations and prevents it from being pushed aside when deadlines get tight.

Knowledge-sharing sessions

Run short talks, lightning rounds, or brown-bag lunches. These are great for spreading expertise across the team. These internal talks are ideal when the budget is tight and your team already has internal experts.

I remember this time when my team had to learn Scala. Our tech lead at the time planned a full week of Scala training, bringing in a Scala trainer from another office to help us. It was a great way to get us started.

I also find daily tech huddles to be useful for encouraging others to share important learnings in a more informal way. These short check-ins after standup give people space to mention discoveries, surface blockers, or discuss small implementation decisions that others might benefit from.

Pair programming or mob programming

These work particularly well when knowledge levels vary across the team.

Pair programming is one of the most effective tools for growth. I've seen it again and again in the teams I've worked with. The strongest learning happens when people sit together and work through real problems, live in the code, sharing context and solutions as they go.

Mob programming extends this concept further, where the entire team works together at the same time, on the same task, using a single computer. While it might seem inefficient at first glance, it can be incredibly powerful for surfacing hidden knowledge and building collective ownership of the codebase, especially in areas where no one has much context.

That said, mob programming tends to work best in very specific situations, like exploring unfamiliar parts of the system or onboarding new team members. It also requires a skilled facilitator to keep everyone engaged and ensure the session stays productive. When done poorly, it can leave people feeling like they're not learning much or that they're not contributing. Like many collaborative practices, its effectiveness comes down to how intentionally it's run.

Hackathons or innovation days

Use these to experiment with new tools or build quick prototypes. I've seen lots of great features come out of hackathon ideas; some even made it to production. As a plus, they boost creativity and are a great team-building exercise.

To make them successful, keep the structure simple: set aside a focused time window (a day or two), let teams self-organize around ideas, and create a low-pressure, playful atmosphere. Encourage demos at the end as a chance to share and celebrate learning.

Post-incident reviews

Postmortems are a great learning opportunity for the team, if you run them right. I dive into how to make the most out of these sessions in the section "Dealing with Incidents" on page 254.

Pairing with other teams

If another team already has experience in a tech you're picking up, ask to shadow or pair with them. You can also explore the option of inviting someone from that team to run a couple of sessions for your group or be available to answer questions. In one case, we even brought a developer into our team temporarily to teach us how to use a specific tool, something none of us had experience with, and it really helped.

When it comes to soft skills, like communication, collaboration, giving and receiving feedback, or even basic Slack etiquette, these are often assumed rather than explicitly taught.

But, just like with tech skills, you can run targeted sessions focused on improving them. I've seen teams benefit from short workshops or informal discussions around things like how to write helpful code reviews, how to have difficult conversations, or how to ask for help effectively. These can be facilitated internally or by bringing in external coaches.

They're a great way to introduce useful concepts and trigger reflection, but real growth doesn't happen in a single session. Soft skills take time, repetition, and consistent practice to build.

That's where process comes in. Processes shape behavior. If your team's routines don't require interaction, interaction won't happen. I've worked with teams where developers barely spoke to each other—no standups, no retros, no shared planning—so no wonder collaboration fell short.

To build soft skills, bake collaboration into your team's day-to-day operations:

- Hold daily standups, weekly planning, and regular retros to create moments for alignment and feedback.

- Encourage shared problem solving through pair programming, whiteboarding sessions, and tech huddles.

- Rotate service ownership to break down silos and build shared responsibility.

- Run regular team rituals that include space for feedback, self-reflection, and improving how the team works together, not just what they build.

- Organize hackathons or "Lunch and Learn" sessions.

You'll find more techniques in the section "Enabling Collaboration Inside the Team" on page 209.

These shared routines help normalize healthy communication habits and give everyone practice in the soft skills they need to grow.

Upskilling a team is about choosing the right tools for your context and making learning part of how your team works every day. Whether it's technical skills or soft skills, the most impactful growth happens when learning is integrated into your team's routines and shaped by what your team actually needs. Be intentional: pick what fits, adapt as you go, and create an environment where people can build new skills through practice, reflection, and shared experience.

TOOLS FOR UPSKILLING INDIVIDUALS

Combine your team-wide efforts with individual growth plans.

Use your one-on-ones to explore each person's aspirations and knowledge gaps, help them build a growth plan (you can help them apply the steps from the section "Developing a Personal Growth Plan" on page 38), track how they're progressing, and help them connect their growth with upcoming work.

Feedback is one of the most powerful tools we have for growth. It shows us what we're doing well, where we need to improve, and how others experience our work.

Here is how you can use it to help individuals develop their technical skills:

Getting feedback on pull requests

One common example is through PR reviews. Developers can track their improvement by the kind of feedback they receive: Are the comments mostly about syntax and formatting, or are they shifting toward deeper questions about design and performance? Are PRs getting approved faster over time, with fewer revisions? These signals can reveal a lot about a developer's growth and confidence in their work.

Getting feedback on technical solutions

Another great opportunity for technical feedback is during solution design.

If someone wants to improve their architectural thinking, they can take the lead on proposing a solution for a new feature. Many teams use shared documents for this; sometimes called TDs (technical designs), RFCs (requests for comments), or architecture proposals. The idea is to write down the plan and share it with relevant stakeholders: team members, other teams, architects, senior engineers.

The feedback that follows is incredibly valuable. It can reveal edge cases they hadn't considered, surface technical risks, or even validate their thinking. Either way, it's a learning moment, and a clear signal of how their current skills are perceived by others.

Once you've supported your team in growing their technical expertise, it's time to focus on the other side of the equation: soft skills. These are the skills that turn strong developers into great collaborators, mentors, and future leaders.

I've been helping techies develop their soft skills for years. I'm an engineer at heart, so I approach soft-skill development just like any technical growth problem: define what to improve, identify the steps to improve it, test something new, reflect on what worked and what didn't, and adapt your process and behavior to include the things that worked.

I've used this process time and again, both as a coach and as a tech lead, to help engineers grow. And it's a process you can apply too, for your own development or to support your team as they build their soft skills.

In the upcoming sections, I break it down, step by step.

Start

You don't need a perfect starting point, just a real one. Look around your team or your current collaboration challenges and pick one thing that's not working as well as you'd like. Maybe you're struggling to give constructive feedback. Maybe you're often misunderstood in meetings.

Pick one problem. Then define a small win. For example, if you want to become better at giving feedback, make a commitment like "This week, I'll give one piece of constructive feedback to a team member." If your goal is to improve meeting facilitation, try "In the next planning session, I'll have a clear structure and keep people on track."

Define what success looks like by defining your "why."

You might think the *why* is obvious, but after coaching hundreds of tech professionals, I can tell you it's not always that straightforward. We all have different motivations. And your motivation, the why behind your goal, will shape how you define success.

Let's take the example "I want to become better at giving feedback."

Now ask yourself, why? Is it because your manager told you to? Is it because people don't seem to take your feedback seriously? Is it because you've noticed things you want to bring up but don't know how? Is it because giving feedback is part of your job expectations?

Each of these whys points to a different version of success:

- If your manager told you, success might mean getting a "meets expectations" rating on feedback in your next review.

- If people don't take your feedback seriously, success could be having someone act on your input or ask follow-up questions.

- If you want to speak up more, success might be delivering one clear, constructive piece of feedback to a team member next week.

- If it's part of your role, success could be getting input from your team about how they perceive your feedback skills.

As you can see, your why determines what progress looks like, and that clarity will guide your next step: taking action.

Apply one change

This is the uncomfortable part. Get out of your comfort zone and try something new to address the problem.

If you usually avoid speaking in meetings, maybe your one change is preparing a point ahead of time and saying it early.

If you struggle with giving feedback, you might try using a framework like SBI and drafting your message in advance.

Not sure what change to make? Here is how you can get some ideas:

Explore resources on the topic of interest

Read a book, take a course, or search online for how others have tackled similar challenges. Even a short article or video can give you a helpful idea.

Use what's available

If your company offers a development budget, make use of it, for conferences, external training, or coaching.

Talk to someone

Reach out to a trusted mentor or coach. They can help you think through the challenge, spot patterns you may not see, and suggest approaches that fit your style.

Observe others

Look around. Who's great at the skill you're working on? What do they do differently? Try borrowing one of their habits or approaches and see how it fits your style.

Progress often comes from small, repeated steps.

Reflect

What worked? What didn't?

This is a key step that often gets skipped in the process of growth. People get caught up in the doing, rushing from one challenge to the next, often repeating the same patterns without realizing it. Reflection makes the growth process intentional. It's how you avoid running in circles, repeating mistakes, and missing out on key insights. It helps you spot what worked and double down on it, amplifying growth far more effectively than focusing only on fixing what went wrong.

To actually understand what worked and what didn't, self-reflection alone isn't enough. You need external input. Soft skills are about human interaction, so the most accurate way to assess them is through feedback: how others experience the change. Any behavior shift will affect the people around you, and their perspective is the most valuable data you can get.

Ask for feedback consistently, especially after moments where soft skills are in play. For example, if you're working on speaking up more in meetings and you share an idea, follow up with a colleague: "How clear was I in that point I made earlier?" The best time to ask is right after the moment happens, when the experience is still fresh and the insight is sharp.

Adapt

Integrate what worked into your habits. Let go of what didn't.

Of course, this is easier said than done. Really incorporating a new behavior takes effort. When we're under stress, we tend to fall back on familiar habits, even if they're not helpful. That's why it requires intentional repetition, preparation ahead of time, and a steady commitment to making the new behavior stick.

Tracking your progress is what keeps the motivation alive. Noticing small wins along the way helps you see that you're moving forward. Mark every step, big or small. Even discovering what doesn't work gives you direction. That insight alone is valuable because it tells you what to avoid and where to focus next. Reflection is about identifying what's working so you can do more of it. In fact, doubling down on what works can have a bigger impact on your growth than focusing only on fixing what's broken. And, as a plus, you're also creating a list of accomplishments and learnings, helpful both when you're feeling stuck and for future reference, like proving growth during performance reviews.

It's easy to get caught in the loop of constant improvement without recognizing progress. If you focus only on what you haven't mastered yet, you risk losing motivation. Instead, regularly recognize what is working. That's what helps you stay energized and engaged, instead of burning out or giving up midway through the journey.

The process of growth is continuous and cyclical. Once the steps—start, apply, reflect, adapt—have been completed, it's time to go through them again. The same situation can be revisited with a new strategy, or a completely different challenge can be tackled.

The great thing about improving a soft skill is that the benefits ripple across everything else you do. You can't isolate one behavior; working on something like listening, for example, inevitably improves how you give feedback, how you build trust, and how you support your team members. As people grow in one area, their overall ability to collaborate, lead, and communicate improves too.

And the impact isn't just individual. When more people on your team level up their soft skills, the entire team dynamic shifts. Communication becomes clearer, misunderstandings happen less often, and conflict becomes easier to manage, or avoided entirely. It's not just about being "nicer" or more "professional." Strong soft skills lead to stronger collaboration and better results.

Some strategies are universally useful, whether someone is deepening their technical expertise or working on communication, leadership, or collaboration. These are great tools to have in your tech lead toolbox:

Encourage people to make use of their development budget, if one is available
You'd be surprised how often this budget goes unused. Many engineers avoid it either because the process seems too complicated or they're unsure how best to spend it.

Help demystify it: walk them through the steps, show them how to get approvals, and advocate for carving time out of work to use it.

Suggest learning options based on their current goals. Conferences, online courses, books, or hands-on learning platforms can all be relevant. Don't limit this to technical domains; make space for things like public speaking, influencing, or writing as well.

Encourage them to work with a coach that can help them reflect, handle feedback better, improve conflict resolution, and work through complex interpersonal situations. It's especially valuable when someone is stepping into a bigger role or facing a major challenge.

Make sure these learning activities are considered part of work time. If they're not, advocate for it: negotiate both the budget and the time off needed to attend events or complete learning programs.

Make learning a shared activity

Amplify growth by turning individual learning into team momentum. Book clubs, Slack channels for sharing course notes, or even show-and-tell sessions during retros or team meetings can help normalize learning in your team culture.

Delegate for development

Delegation is a great growth tool. Chapter 6 explores how to use it effectively. For example, if you always handle a specific design review or infra setup, consider pairing with a team member and handing it off, guiding them while they learn.

Encourage peer or external mentorship

If someone wants to grow in a specific technical or leadership area, help them find a mentor. This could be someone within your company or through a professional mentoring platform.

Helping someone grow means offering direction, tools, and support while creating space for learning to take root. Your job isn't to do the learning for them but to make it easier for them to move forward with clarity and confidence. Encourage ownership and initiative while also removing friction where you can. With the right structure, your team can take charge of their development and keep making progress.

LEARNING TRAPS TO HELP YOUR TEAM AVOID

Growth, whether technical or interpersonal, is rarely linear. Even with the right intentions, people often run into common traps that slow or derail progress. As a tech lead, your role is to help people recognize these pitfalls and navigate around them:

Trap: "There's no time to learn"

We're all busy all the time; the real question is this: are we busy with the right things? Learning won't happen unless you make space for it. That means reprioritizing. Some things might need to drop. But if growth is truly important—and it is—there's always a way.

Trap: Consuming without applying

Sometimes, we find ourselves doing things but not progressing. You can read all the books and take all the courses in the world, but if you don't apply what you've learned, it doesn't stick. I've seen this loop many times. Encourage your team to turn learning into practice.

For example, if someone is learning about systems design, give them ownership of a relevant architectural discussion or RFC.

Or, if someone wants to improve their facilitation or communication skills, they could lead a team initiative, facilitate a sprint retro, or represent the team in a cross-functional meeting.

Trap: Learning in isolation from business needs

Learning topics can be disconnected from the team's or business's priorities. Growth should be intentional and useful. Help your team connect what they're learning with what the business actually needs. Encourage your team to ask, "How will this skill contribute to the team/product?" Help connect learning efforts to real outcomes.

Trap: Goals that are too big or too many

People often try to fix everything at once. The problem with this trap is that when goals are too big or too numerous, they become overwhelming and demotivating. People either burn out trying to tackle everything at once or fail to start altogether because it feels unmanageable.

Encourage them to start small and focus on tangible outcomes. For example, instead of aiming for "All my team members give me a lot of useful constructive feedback in the next month," help them redefine it to "One team member gives me one piece of constructive feedback in the next two weeks."

Or if the challenge is around decision making, help them go from "Everyone agrees and is happy with the solution" to "We agree on the problem we're trying to solve" or "We move forward even if not everyone agrees, but we have a shared commitment to the decision."

Small wins build momentum and confidence.

Trap: Skipping reflection altogether or relying just on gut feeling to know if they're improving

This is one of the most common traps, and it's easy to fall into, especially when you're busy or focused on outcomes. People keep trying new things but without taking a moment to step back and ask, Did that actually work? What impact did it have? Without reflection, there's no clarity, no momentum, and no intentional learning.

Instead, help your team build a habit of regular reflection and feedback.

For example, since soft skills are all about how we affect others, the best way to understand progress is to ask for input. Encourage your team to request feedback after key moments like meetings, presentations, and decisions, especially when they've tried something new.

Trap: No accountability

Growth takes time and consistency. Without a system of accountability, progress often fades. That's where you come in. Be their accountability partner, or help them find one, whether it's a mentor, a peer, or even a coach. Regular check-ins make learning stick.

Growth takes intention. Help your team stay focused, avoid common pitfalls, and keep learning as part of the everyday workflow. Small nudges, regular check-ins, and space to reflect can make all the difference.

Managing Stakeholders

As a tech lead, your impact doesn't stop at the edge of your team. You also have to manage the relationships that surround your team, the people who influence what gets built, how work gets prioritized, and whether your team's work is truly understood and supported. That's where stakeholder management comes in.

In this section, I'll explore why managing stakeholders is a core part of your role.

You'll learn how to identify the right stakeholders, understand what they care about, and build relationships that protect your team and amplify your work. From mapping influence to navigating difficult conversations, I'll walk through practical techniques that help you manage these relationships with clarity, empathy, and intent.

WHY YOU NEED TO MANAGE YOUR STAKEHOLDERS

As a tech lead, you are the bridge between your team and the rest of the business (Figure 10-1).

Stakeholders rely on you. You're their connection to what's happening day-to-day. When things are going well, it's on you to make that visible. When there are risks, blockers, or changes in delivery, you're expected to surface those early and clearly explain the impact. This is where your role goes beyond planning and execution; you're also shaping the narrative between what your team is doing and what the business expects.

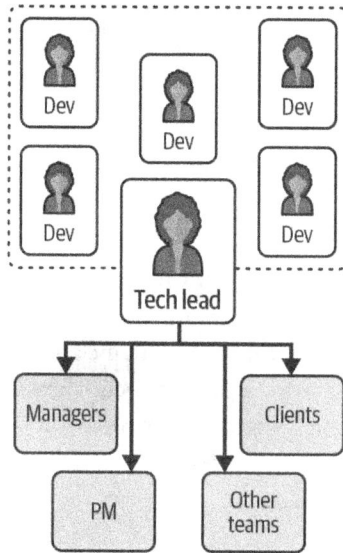

Figure 10-1. Tech lead as a bridge

This part of the role doesn't happen automatically. You can't wait for stakeholders to come to you. Managing stakeholders well takes intention and consistent effort. It starts with building strong relationships, through regular one-on-ones, ongoing updates, and shared systems that make the team's work visible. These relationships should be about building trust, aligning context, and helping stakeholders feel confident in the work being done.

Start with your product manager. While both tech leads and product managers engage with stakeholders, their focus differs. PMs typically lead on defining the "what," the product vision, requirements, and prioritization, while tech leads focus on the "how," translating those requirements into feasible plans and identifying technical trade-offs. But this division works only when the partnership is strong. Together, you form the team's core leadership, and how well you collaborate shapes everything from delivery speed to team morale.

So set up recurring one-on-ones. Align early on how you'll communicate, make trade-offs, and resolve disagreements. When the relationship is healthy, your collaboration brings clarity, consistency, and stability. When it's weak, the team feels the tension, in the form of mixed messages, shifting priorities, or unresolved conflict. Like any partnership, this one takes care and intention. Invest in it.

And the same goes for your other stakeholders. When these relationships aren't prioritized, misalignment becomes almost inevitable. And when alignment breaks down, execution suffers, no matter how technically solid the work is.

Almost every month, I hear from a tech lead who's frustrated that their stakeholders are disappointed. Delivery came late, the solution missed the mark, or it wasn't what was expected. And the pattern is often the same: the tech lead focused on execution, assumed alignment, and only brought stakeholders in late, when it was too late to course-correct. These conversations were seen as a distraction from "actual work," when in reality, those conversations are part of the actual work. If you skip early alignment, you might build something technically perfect and still fail.

I'll dive into techniques for avoiding this scenario in the upcoming section "Stakeholder Management Techniques" on page 323. But it starts with a mindset shift: managing stakeholder relationships is not separate from your work. It is your work.

At the same time, your team needs you to bring information back to them. Stakeholders might change timelines, priorities, or expectations. You need to translate those shifts into clear, actionable guidance. Sometimes that means shielding your team from noise or unrealistic demands. Other times, it means helping them see the bigger picture.

Let's say you've just learned that a company-wide strategy shift is coming. Your team is mid-way through a six-month plan, but priorities are changing because a key third-party service is increasing prices. The business no longer wants to pay for it and is requesting a migration to a cheaper provider within two months. You now need to go back to the team and develop a plan.

They won't like the news, and how you deliver it will determine how they respond. If you can explain why the change is happening, share the reason behind the tight timeline, propose an alternative provider, and secure stakeholder buy-in to adjust the scope up front, your team is more likely to get on board.

Anticipating their questions—Why this change? Why now? Will this mean extra hours?—allows you to lead with clarity and empathy.

You don't need to have all the answers before talking to your team. You'll make the plan together. But the more context you gather ahead of time, the smoother that conversation will be. Go in with the attitude of "I know this isn't ideal, and I need your help to figure it out," rather than "This is what we're doing."

Finding that balance—being the translator, the advocate, and the filter—is what makes this part of the role so critical.

It's also important to continuously keep in mind that relationships are always two-way. Just as stakeholders are essential to your team's success, your team is essential to theirs, and they know it. This means you shouldn't view stakeholders only as people in charge or gatekeepers but as allies who share a common goal. Ask for their help when you need it. If there's a blocker, ask them to use their influence to help clear the way. If a decision needs broader buy-in, ask them to support you in championing your proposal. They're not just there to be managed; they're there to help, too. It's in both your interests to succeed. Frame the relationship as a partnership rather than a hierarchy.

I know it can be frustrating, but success isn't just about what you deliver; it's about what's recognized. You could build the most technically impressive product in the company, but if no one knows about it, it doesn't move the needle. Like it or not, marketing matters. It's your job to make your team's impact visible by tying outcomes to business value and communicating in ways stakeholders will hear and understand.

Managing stakeholders is one of the core ways you create impact as a tech lead. Do it well, and your team will have the clarity, trust, and support it needs to thrive.

IDENTIFYING YOUR STAKEHOLDERS

As the tech lead, you are the voice and shield of your team. You're the one with the full context of what the team is building, how they're working, and what challenges they face. It's your job to represent the team's needs and interests with clarity and accuracy, to shield the team from distractions or misaligned external pressures.

To effectively represent your team, you need a clear view of the larger organizational context. You need to understand how the company is structured, who holds decision-making power, who controls resources, and who might be unexpectedly relevant.

A great way to begin this process is by reviewing any available documentation on your company's organizational structure. Many companies maintain visible org charts outlining roles and hierarchies.

If this structure doesn't exist, create your own. Start by listing everyone who might influence your team's work. Stakeholders exist at multiple levels, not just senior leader or formal roles. Consider collaborators from across the organization, such as product managers, designers, QA, infrastructure, and security teams, as well as adjacent product teams with shared dependencies.

Also be aware of informal influence. Influence isn't always tied to job titles. Some people shape direction through relationships, expertise, or institutional knowledge. Think of long-tenured engineers, well-connected peers, or customer-facing staff who engage regularly with users. For example, there is often a developer who's been with the company for a long time—maybe even from the beginning—or someone who built a core part of the product. Even without a formal leadership title, people look to these people for guidance and opinions. That's definitely someone to add to your stakeholder list.

Pay attention to who others defer to in meetings, who helps resolve cross-team issues, or who consistently brings hidden context to light. These individuals may not appear on an org chart as decision makers, but their impact is real and often significant. Including them in your stakeholder strategy can make the difference between reactive firefighting and proactive alignment.

As you build this list, consider the following questions:

- Who can significantly affect my team's ability to succeed?

- Who could amplify our success if I had a stronger relationship with them?

- Who is currently making decisions that affect my team, even indirectly?

- Who holds information, context, or resources that could help us succeed?

- Who might block our work, intentionally or not, if we don't engage them early?

- Who has a vested interest in the outcomes we are working toward?

These questions will help you identify which relationships to invest in and guide your strategy for building and managing stakeholder connections.

If you're uncertain about who truly holds influence or how decisions are made, talk to people. Short, informal one-on-one conversations can be incredibly revealing. Ask questions like "Who else should I be talking to?" or "Who influences decisions around this area?"

As you uncover more about your stakeholders, consider documenting these insights. Keep a lightweight, private record that captures who your key players are, what they care about, how they prefer to communicate, and any notable context or risks. Such a resource becomes especially valuable when onboarding new leads, sharing knowledge with peers, or navigating shifting team dynamics.

One common tool for visualizing your stakeholders is the Mendelow Power-Interest Matrix, which helps you map stakeholders based on how much power they hold and how much interest they have in your team's work. It's a simple way to guide your engagement strategy—who to involve closely, who to keep satisfied, and who just needs occasional updates.

Another useful tool, and the one I'll explore in the next section, is the stakeholders map. It uses a straightforward two-axis framework: one axis for influence (meaning how much they can impact your team) and one for commitment or alignment (whether they are a supporter or a detractor).

Intent also exists on a spectrum. Supporters believe in your goals and work with you to achieve them. They are your partners in success. Detractors may not yet believe in your work or might even oppose it. Often, this opposition is not personal; it may come from broader organizational dynamics or differing priorities.

STAKEHOLDER MANAGEMENT TECHNIQUES

Now that you have your stakeholders mapped out, it's time to put that map to use. The goal here is to tailor your engagement strategy based on where each stakeholder sits in terms of their influence and commitment.

The map (Figure 10-2) gives you a clear picture of where to focus your attention.

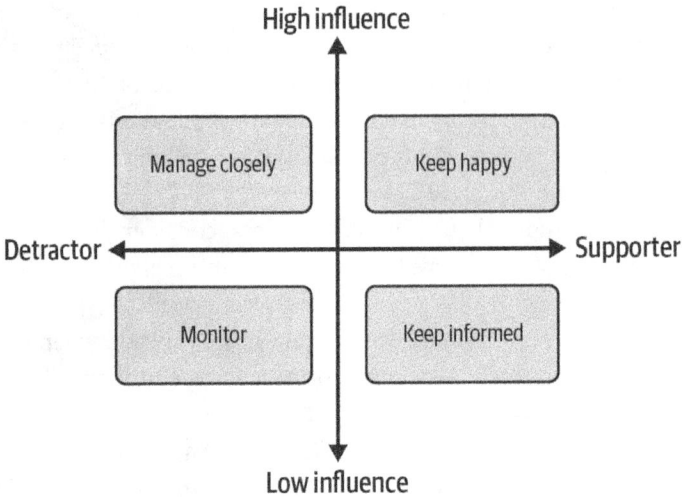

Figure 10-2. Stakeholders map

Stakeholders generally fall into four categories:

- High-influence detractors are critical to manage closely. These individuals can block or derail progress if not engaged carefully. Understanding their concerns and working to build alignment is essential.

- High-influence supporters are your champions. Keep them happy, informed, aligned, and engaged. Their advocacy is powerful.

- Low-influence detractors should be monitored. They don't require heavy investment unless their influence grows.

- Low-influence supporters should be kept in the loop. They may grow in influence or help provide useful context and feedback.

Empathy is key

Empathy is key, especially when working with detractors. What's driving their perspective? Are they under pressure themselves? What do they want to see happen?

For example, while working in a consultancy, we were preparing to roll off a client project and needed to hand over our work to their internal team. We proposed pairing on tasks, a method we believed would transfer knowledge quickly while still delivering value. But an engineering leader on the client side was strongly against it. Despite offering every rationale—speed, quality, continuity—he wouldn't budge.

So I teamed up with the account manager and scheduled a call with him. We didn't try to convince him again. Instead, we listened. He shared concerns based on past experiences: handovers had taken too long, the team lacked experience, and remote collaboration tools weren't ideal. He was also nervous about how the change would land with his team.

Only after fully hearing him out did we respond. At this point he seemed way more open to conversation. We explained how we planned to address those concerns: checking in with his engineers to ensure they were also on board, building a lightweight plan with clear support, and committing to a short trial period to see how it went. He agreed to those actions.

The team got on board, and the pairing experiment worked surprisingly well. We kept him in the loop at every stage, shared updates, and encouraged feedback from the engineers on the ground. By the end of two weeks, he was all in for pairing as a handover approach.

That experience taught me that when resistance shows up, your first job is to understand. Behind most pushback is a story: a past experience, a pressure, a concern that hasn't been voiced yet. Once we acknowledged his concerns, reflected them back, and offered a low-risk way to move forward, trust started to build.

Sometimes, people just need to feel heard before they can move forward. Other times, they need help navigating internal optics, such as how a decision might reflect on them within their own team. Being mindful of that, and helping them look good in front of their peers, can make all the difference.

And sometimes, the most powerful thing you can do is remind them that you're on the same side. Reassure them that you care about finding a solution that works for everyone. Simple phrases like "I want to make this work for all of us" or "I need your help to figure this out" can lower defensiveness and invite collaboration.

In the end, empathy turned that blocker into a collaborator. That's the power of showing up curious instead of defensive. And it's one of the most valuable tools you have in stakeholder management.

Adapt your communication style

A big part of stakeholder alignment comes down to communication. One helpful framework for adapting your style to theirs is the DISC model, originally developed by psychologist William Moulton Marston, which classifies behavior into four primary styles (Figure 10-3):

Dominance (D)
> Results-focused, assertive, challenge-driven

Influence (I)
> People-oriented, enthusiastic, social

Steadiness (S)
> Calm, consistent, team-oriented

Conscientiousness (C)
> Precise, analytical, data-driven

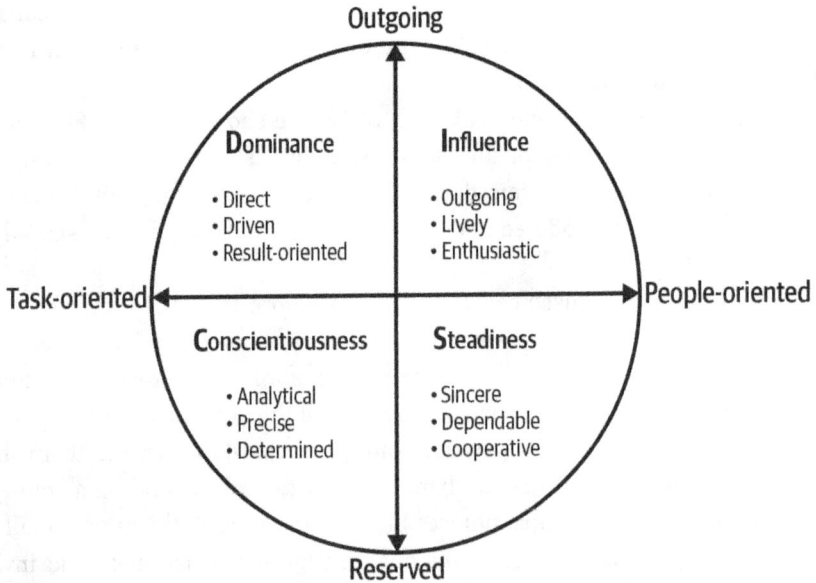

Figure 10-3. DISC model

Understanding where your stakeholder falls on this spectrum helps you communicate more effectively:

- For D styles, be direct and focus on solutions. Let them make decisions.
- For I styles, engage with enthusiasm, share stories, and allow them to ask questions.

- For S styles, be personal and friendly, give them time to process, and avoid confrontation.

- For C styles, stick to facts, reduce emotional language, and allow time for them to process the information before giving a response.

It's equally important to understand your own default communication tendencies and how they might clash with others. DISC styles that sit opposite each other on the spectrum tend to experience the most friction.

For example, I naturally fall into the C (conscientiousness) style; I like structure, precision, and thoughtful communication. But I've struggled at times when working with I (influence) stakeholders. Their fast pace, high energy, and preference for informal, emotional connection can feel overwhelming or vague to me. Meanwhile, they may find my style too rigid or cautious.

Recognizing these gaps has helped me adjust. With I styles, I've learned to loosen up, speak more casually, and share more personal context. That small shift in tone often opens the door to stronger connection and trust.

Adjusting your style means being intentional about how you show up so you can be heard instead of ignored.

Build trust over time

Adapting your style is one way to build trust, but nothing beats the kind that comes from time and consistency. That's why my favorite strategy for stakeholder management is building trust over time. It's not always possible; some environments move too fast or change too often for deep relationships to form. It requires continuity and time spent working closely with the same stakeholders. But if you do get that opportunity, take it. The payoff is worth every bit of effort.

I was lucky to be in environments where this kind of consistency was possible—same team, same stakeholders over a long stretch of time. That stability made it easier to invest in strong relationships and see the benefits. I encountered relatively little resistance because trust was already there. When my team had to make a tough technical call that delayed delivery, the stakeholder's response was simply "If you say it's needed, it's needed. I trust you." That kind of response is earned through how you consistently show up.

If you're in a stable context, I highly recommend applying the principles from the section "How to Build Strong Relationships" on page 62. Focus on consistent and transparent communication: hold recurring one-on-ones with stakeholders, go beyond just sharing updates, communicate risks early, seek and incorporate feedback, and most importantly, follow through on your promises. Delivering consistently builds credibility.

Strong collaboration comes from trust, consistency, and shared ownership.

So when you do get the chance to invest in long-term relationships, make it count. They'll pay off far beyond any single project.

A powerful but often overlooked tip: involve your team in the process of building stakeholder relationships. Teach your team members why these relationships matter and how to build them. For example, if you have a feature lead running an initiative that depends on close collaboration with another team, encourage them to set up recurring one-on-ones with the lead on the other team, create shared communication rituals, and celebrate wins together. When more people on your team build strong relationships across the organization, your collective impact grows, and so does yours, by extension.

Bringing it all together: A lightweight framework

To make these strategies easier to remember and apply, I use a lightweight framework I call the 3Rs of stakeholder alignment:

Reframe
Translate technical needs into business value.

Relate
Connect the work to stakeholder goals and priorities.

Reinforce
Share progress and outcomes consistently to build confidence.

This lightweight mental model can serve as a quick checkpoint anytime you prepare to engage stakeholders about a decision.

Stakeholder alignment is about building trust, finding common ground, and creating momentum together. Like any leadership skill, it gets stronger with practice, reflection, and consistency. The more you invest in these relationships, the easier it becomes to navigate complexity.

Common Challenges in Navigating Technical Leadership

Even experienced tech leads can struggle with the realities of the role as it requires navigating an environment full of competing priorities, shifting expectations, and limited time. This section explores some of the most common (and often unspoken) challenges you'll face in the day-to-day of technical leadership.

I'll look at how to secure stakeholder buy-in for technical initiatives, especially when the value isn't immediately visible. I'll also explore how to measure your own success in a role that often lacks clear metrics. And finally, I'll confront the emotional and mental toll of trying to do it all, and what it really takes to lead without burning out.

GETTING STAKEHOLDER BUY-IN ON TECHNICAL TOPICS

Securing time from stakeholders to focus on technical topics often proves challenging for tech leads. Whether it's addressing tech debt, resolving deep system issues like memory leaks, or improving internal tools for better development workflows, advocating for technical work can often feel like an uphill battle.

The core problem is that many tech leads approach these conversations like a confrontation. They brace for resistance and forget a crucial truth: stakeholders and engineers are on the same side. They all want the same outcome: successful delivery and sustainable progress.

Often, when stakeholders don't give their buy-in, it's not because they're against the work; it's because they don't fully understand the value, urgency, or impact. The misalignment in priorities and language can make technical work seem abstract, low priority, or disconnected from business goals. The key is finding that common ground, and that's where the 3Rs of stakeholder alignment comes in (described in the section "Stakeholder Management Techniques" on page 323).

Here's the framework applied to some of the most common technical topics tech leads need to advocate for:

Addressing an underlying system issue
> One memorable example for me was a persistent memory leak in one of our most critical services, built on .NET. Initially, the problem appeared minor: the service would occasionally stop responding. Restarting the servers fixed it temporarily, so that became our short-term workaround.
>
> We were busy launching a new project, and this recurring issue kept getting pushed aside. Eventually, a new team member joined and asked the hard question "Why is this happening again and again?," so we could no longer ignore it.
>
> We proposed a one-day spike to investigate, which quickly uncovered a serious memory leak. Fixing it wasn't going to be fast, and it wasn't something we could afford to ignore either. I needed product buy-in to dedicate the time.

I got that buy-in by applying the 3Rs of stakeholder alignment framework:

Reframe

I explained what the memory leak meant, not just for the business: increased infrastructure costs, service downtime, and poor user experience.

Relate

I tied the issue to broader stakeholder concerns, including customer impact and potential long-term delays.

Reinforce

I committed to providing regular updates via standups and sprint reviews, which helped maintain transparency and confidence.

Once the issue was fixed, the job wasn't over. We measured and shared the impact: how much time was saved, how reliability improved, and how morale lifted. Mentioning these results in retros and sprint reviews helped stakeholders see the return on their investment.

Advocating for a major refactoring

A common situation many tech leads face is advocating for a major refactoring of a legacy system. From the outside, the system may appear stable, but internally it's brittle, difficult to maintain, and slows down the team's ability to ship new features with confidence.

To secure stakeholder buy-in for a large refactor, you can apply the 3Rs of stakeholder alignment:

Reframe

Explain that while the system is still functional, it's causing delivery slowdowns and increasing the likelihood of bugs. Position the refactoring not as internal cleanup but as an investment to speed up feature development and reduce long-term risk.

Relate

Connect the challenges your team is facing with recent stakeholder pain, like missed deadlines, escalating QA time, or inability to capitalize on new business opportunities. Be explicit about timelines and impact: propose a clear plan that outlines how the refactor affects the current roadmap and how you're mitigating disruption.

A piece of advice here: steer clear of proposing a months-long "let's stop everything and rewrite the whole thing" plan. The cost and business risk are too high, and your stakeholders will rightfully push back on pausing product progress for months. Instead, look for ways to break down the work, tackle it incrementally, and continue delivering value throughout.

Reinforce

Commit to visibility. Share progress through sprint reviews or status updates, and highlight the gains with before-and-after comparisons.

Dealing with tech debt

You can find the 3Rs of stakeholder alignment framework applied to advocating for tech debt in the section "Managing Technical Debt" on page 249.

Getting stakeholder buy-in on technical topics doesn't have to be a battle. When you apply the 3 Rs—reframe, relate, and reinforce—you shift the conversation from resistance to collaboration.

Of course, buy-in doesn't always mean getting everything you ask for. Sometimes it involves negotiation, compromise, or finding middle ground, perhaps reducing the scope, tying the work to a business milestone, or deferring part of it. These trade-offs are a normal and necessary part of the process. Just as important is knowing when to push and when to let go.

For example, you might propose rebuilding a fragile internal tool, but your stakeholders push back due to an upcoming product launch. Rather than insisting on doing it all at once, you apply a negotiation mindset and look for mutual gain. You agree to reduce the scope and focus on just the most unstable part of the tool now, which helps avoid potential outages during the launch, and defer the rest to a quieter release window.

Even if you do everything right, sometimes it just doesn't go your way. You might communicate clearly, align goals, and build trust, and still hear "no." For example, you might propose introducing automated performance testing to catch regressions early. But stakeholders decline because they don't see an immediate need, especially with a major product launch approaching. Rather than walking away frustrated, you follow up to understand the reasoning. You learn their concern is about adding complexity during a high-stakes window.

Instead of forcing the issue, you agree to revisit the proposal after the launch. In the meantime, you begin collecting examples where performance regressions slipped through unnoticed, helping to build a stronger case for the future. This approach keeps the relationship intact, demonstrates patience and strategic thinking, and lays the groundwork for future buy-in.

With these strategies in place, you can advocate effectively for your team's needs while keeping stakeholders engaged and supportive.

As a next step, reflect on this: what's one technical priority your team needs stakeholder support for right now? How could you reframe it in terms of business value? Who do you need to start building a relationship with today to make that conversation easier tomorrow?

MEASURING SUCCESS IN TECHNICAL LEADERSHIP

Success in technical leadership isn't as clear-cut as shipping features or fixing bugs. It's about influence, alignment, and long-term impact: things that don't show up in commit histories or sprint velocity charts. And that makes it incredibly hard to measure.

As a tech lead, your role sits in a blurry middle ground: not quite management, not purely an individual contributor. You guide architecture, support delivery, grow team members, and bridge communication between technical and nontechnical teams. But there's often no formal rubric that tells you how well you're doing.

That's what makes this so tricky: you're expected to lead but rarely told how success will be judged. And when there's no clear feedback loop, many tech leads start to second-guess themselves. They wonder if they're doing enough. Or worse, if their contributions even count.

The ambiguity leads to a lot of quiet frustration. Performance evaluations may still rely on outdated metrics like lines of code, Jira tickets closed, or how "busy" you appear. These don't reflect the real, valuable work of unblocking others, maintaining technical quality under pressure, or shaping a healthy engineering culture.

Meanwhile, your most impactful contributions, like preventing system failures or scaling team processes, might go completely unnoticed unless you surface them intentionally.

The simplest way to assess your progress as a tech lead is by measuring yourself against the goals you've set. These goals act as your compass; they help define what success looks like in the short term and give you something concrete to aim for.

Setting these goals should be part of your broader growth plan. You can read all about how to develop one in the section "Developing a Personal Growth Plan" on page 38.

Set one or two focused goals each quarter based on what your team or organization needs most. For example:

- Improve team visibility by launching a monthly showcase for stakeholders.

- Strengthen cross-team communication by setting up a recurring sync with adjacent product teams.

- Expand our impact by taking ownership of the internal authentication service and driving its roadmap.

These goals should connect back to your team. After all, your success as a tech lead is inseparable from how well your team is doing. Set goals that reflect their growth, unblock their challenges, and support their performance. The clearest signal that you're on the right path? A thriving, confident, high-performing team. These goals may feel outside your control at first, but that's part of stepping into leadership. The good news is: you have more influence than you think.

Revisit these goals regularly. Ask yourself: Have I made visible progress? Have I unblocked or accelerated anything as a result? What feedback have I received that supports or challenges this progress?

Anchoring your growth to these clearly defined goals not only helps you reflect and course-correct along the way; it also gives you a practical, proactive way to demonstrate your value as a tech lead: with measurable progress rather than abstract qualities.

Another useful way to assess your impact as a leader is to pay attention to your gut feeling. Are you consistently hearing positive feedback, or are there recurring concerns about certain aspects of your leadership? Do people come to you for guidance and decision making? If not, what might be preventing that? These are powerful indicators of how you're perceived and whether your leadership is being felt.

Your gut instinct can offer valuable signals, but it's just one input. To validate it, you need to seek out direct feedback. To get a more complete picture, gather feedback from all directions, for both yourself and your team. This means team members, stakeholders, partner teams, cross-functional collaborators, even clients or users.

You can start small. Sometimes all it takes is a casual check-in: "Hey, I've been thinking about how we've been working together lately. Is there anything I could do to make collaboration easier?"

Or you can take a more structured approach in the form of a longer conversation. Use open-ended questions that encourage honest input and help surface where your impact is most visible. Here are some examples:

- What kind of impact do you think I've had on the team as a tech lead?
- How well do you think that impact aligns with what you expect from someone in this role?
- What are the areas where you feel I didn't contribute as much as I could have?
- What would you suggest I focus on to improve?
- What should I start doing differently?
- How would you describe your experience working alongside me?
- What should I be focusing on right now that I may be overlooking?

Whenever possible, tie these questions back to the goals you've set. For example, if your focus is on improving collaboration, you might ask: "How effective do you think I've been in bringing people together to solve problems?"

The key step is to track this feedback. Make it a habit to write down what you hear, including who said it, when, and in what context (e.g., during a critical incident, a big release, or a high-stakes delivery). Keeping this kind of record helps you identify patterns, document impact, and, most importantly, prove your value as a tech lead. When it comes time for performance discussions, this evidence will help you clearly articulate the influence you've had—not just what you did but the outcomes your leadership helped create.

Another powerful metric to assess your effectiveness as a tech lead is overall team performance. When the team is performing well, shipping high-quality work, collaborating effectively, and growing in confidence, it's a clear sign of strong leadership. On the other hand, if the team is struggling, it's often a reflection of gaps in leadership, support, or clarity. That's why it's essential to continuously assess not just your individual impact but your team's performance as a whole. There are structured ways to evaluate how well your team is functioning. You can explore those in the section "How to Evaluate If Your Team Is High-Performing" on page 195.

Whether sprint demos or milestone reviews, your showcase rituals are simple, visible signals of team effectiveness. These sessions are more than routine ceremonies; they provide a track record of your team's results and outcomes.

Are you consistently demonstrating progress? Is the work tied clearly to business value? Regular, high-quality showcases reflect alignment, focus, and momentum, all of which point back to strong technical leadership.

Another great question to ask yourself is this: how is your team doing when you are not there? It's actually a sign of great leadership if your team is properly functioning even in your absence. A team that can operate effectively without your constant presence is a strong indicator that you've empowered them well.

Understanding your own impact, and helping others understand it too, allows you to focus your energy where it matters, advocate for the recognition you deserve, and grow with confidence in your leadership journey.

BALANCING THE DEMANDS OF TECH LEADERSHIP TO AVOID BURNOUT

A common trap tech leads fall into is trying to manage it all at the same time. We think we can do it all: write code, lead people, align stakeholders, and handle operations. But without clear boundaries or balance, this can quickly lead to burnout. You may find yourself working late to catch up on what didn't get done during the day or constantly switching contexts, never fully present in one mode or the other.

I often hear tech leads say, "All my real work happens outside of working hours because I'm always trapped in meetings." But those meetings are real work, and they are often harder than any technical task. They're where alignment happens, risks get surfaced, and relationships are built.

Still, it's an antipattern for tech leads to spend all their time in meetings. Not every meeting needs you. Protecting time for deep work, team connection, and strategic thinking is essential. Delegate when you can. Push back on meetings where your presence isn't adding value. Block focus time, and treat it like you would any other critical commitment.

Constantly trying to be everywhere is how that pressure builds, and it slowly leads to exhaustion. Some people notice the signs early. I didn't.

Just five weeks into my first tech lead role, I experienced facial paralysis. I basically couldn't move the right side of my face. I felt bad for my team and for myself. And yet, I took only a couple of days off, even though recovery typically takes months.

I came back to work taking strong medication and wearing sunglasses for weeks so people wouldn't see I couldn't fully close my eye. I even joked about it.

At the time, I thought I was doing the right thing, pushing through, showing strength. But in truth, I was afraid. Afraid of disappointing people, of being seen as weak, of not rising to the challenge. I also realized I saw myself as a bottleneck for my team; I didn't trust them to handle things alone.

I've thought a lot about the message that was sent. I told myself I was being committed, but what I really modeled was self-neglect. I normalized working through serious health issues, and I made it harder for others to feel safe prioritizing their own well-being. That's not the kind of leader I want to be.

Despite ignoring the doctor's advice, I was lucky to recover quickly. But that doesn't mean the impact wasn't real. The stress lingered, and I lost trust in my ability to set healthy boundaries. It took me months to start integrating the lessons from that experience into how I lead.

This is one of the reasons I wrote this book: to help others recognize these patterns earlier. No deadline, sprint, or meeting is more important than your health. Your team needs a leader who's well and who makes it safe for them to be human too.

Since then, I've been more intentional about how I work and lead. Here are some strategies that have helped me and that might help you too:

Keep yourself in check. Constantly.

How are you feeling? How are you managing it all? What do you need?

Ask people around you too. Sometimes we're so trapped in our own heads we miss the signs, but others notice. I remember when I wasn't feeling OK, I started getting more "How are you doing?" or "How's everything?" from people around me. It was like they could sense something. One time, a team member said in a one-on-one, "I feel like you're stressed. What's happening?" It caught me completely off guard; I thought I was hiding it well. Trust those cues.

Ask for help. Early.

The tech lead role can feel lonely, but you are not alone. Admitting you need support and accepting it is essential.

Here's who I've leaned on: my manager, someone I could turn to for advice, direction, or just to listen; other tech leads, through internal groups or one-on-one chats that helped normalize challenges; a coach, someone objective to help me think things through; a mentor inside the company I met with regularly; and my team.

When I admitted I was overwhelmed and showed them my to-do list, they volunteered to help. We found things they were interested in. That act empowered them and was a relief for me.

There is support around you. Sometimes, you just have to reach out.

Take some time off.

Many tech leads feel they can't take a vacation, that their team can't function without them. But if that's true, your team is too dependent on you. Great leadership means your team can perform even when you're not there.

Plan more living, not less working.

I saved my favorite strategy for last because it completely changed how I think about balance.

For a long time, I thought the key to avoiding burnout was working less. So I tried things like starting work later, shutting the laptop at 6 PM, or adding more breaks to my day. But the problem wasn't just the amount of work; it was what I wasn't doing outside of it.

When I stopped working, I found myself wondering: now what? I had no hobbies, no post-work plans, no clear way to transition out of work mode. That emptiness made it easier to just keep working, and harder to disconnect.

Eventually, I realized I needed to plan life the same way I plan work. So I started doing more life. I began booking activities after work that I paid for in advance, giving myself skin in the game so I'd actually follow through. I made plans with friends, creating social commitments I didn't want to cancel. I picked up small personal projects, like building Lego sets or solving puzzles, that gave me the same sense of progress and momentum I enjoy at work. And I got a dog, which naturally forced me to take breaks and spend more time outside.

This is about making space for life to grow around it. If you want to sustain your energy and protect your well-being, schedule living with intention.

Prioritizing your self-care means prioritizing your team's health too. They look to you for cues. If you overwork, they overwork. If you're stressed, they get stressed.

If you're not well, you can't help anyone.

Key Takeaway

Technical leadership is rarely about having all the answers; really, it's about creating the conditions for your team to thrive. In this chapter, you've seen how that comes to life across the areas that matter most: developing talent, managing stakeholders, and bridging the gap between tech and business.

Your job as a tech lead is part translator, part strategist, part coach. You guide technical direction, support the people around you, and navigate relationships that extend far beyond your team. Each decision you make—how you spend your time, how you communicate, how you support growth—sends a signal and shapes your team's culture.

Sustainable, effective leadership doesn't come from doing more. It comes from doing the right things, with clarity and intention. Start where you are. Pay attention to what's working. Ask for help when you need it.

The most impactful tech leads aren't perfect; they're thoughtful, adaptable, and always learning.

Beyond Tech Lead: Charting Your Career Path

If you've been in the tech lead role for a while and you're ready to grow into your next stage, or even if you're just curious about what's possible, this chapter will help you explore your options.

There's no single direction you're supposed to go from here.

Some people choose to stay in the tech lead role a bit longer, perhaps taking on a new challenge: leading a different team, working in a new domain, scaling up to a larger group, or trying it at another company.

Others transition fully into people management, dive deeper into technical leadership, or explore entirely different ways of leading, through programs, strategy, or consulting.

Before you make any big decisions, this chapter invites you to reflect on your experience so far, get honest about what you've loved (and what you haven't), and evaluate your strengths and areas for growth. From there, I'll explore the most common career paths after tech lead and how to approach the transition in a thoughtful, low-pressure way.

Your next step doesn't need to be permanent. It doesn't even need to be perfect. It just needs to move you closer to the kind of work, and the kind of impact, that feels right for you.

Let's dive right in.

Reflecting on Your Experience as a Tech Lead

Before you can figure out where to go next, it's worth pausing to take stock of where you've been.

The tech lead role is rich with learning, about systems, about people, and, often, about yourself. Maybe you stepped into it with curiosity. Maybe it was handed to you without much preparation. However you got here, now's your chance to look back and ask, What has this role taught me? What have I enjoyed most? What has felt draining?

This section is not about evaluation in the performance review sense but about self-awareness. The kind that helps you understand what kind of work brings out your best, what kind of impact you want to have, and what kind of roles might be a better fit going forward.

You don't need perfect answers, just honest ones.

UNDERSTAND WHAT YOU ENJOY (AND DON'T) ABOUT THE ROLE

Before you decide what's next in your career, it's important to pause and reflect: what has this role actually felt like for you?

Stepping into technical leadership is a big shift. You probably entered the role with some expectations, maybe even excitement, about the opportunity to lead, guide, and create impact beyond just the code.

But expectations and reality often don't align. This section is your chance to make sense of that gap. What surprised you? What challenged you? What inspired you?

To begin this reflection, it helps to break it down into a few key areas:

What you've enjoyed most

Start by thinking about what you've genuinely enjoyed. Which parts of the role gave you energy? Maybe it was delegating a task to a team member and watching them grow. Maybe it was working across functions to align on a complex project. Maybe it was having more visibility into business decisions and helping translate them into technical action.

Try to remember the full picture, not just the highlights. It's easy to forget the frustrating parts, like long meetings, messy handovers, endless alignment. Sometimes it's not the task we miss but the outcome. You might think you miss coding, but what you're really chasing is the dopamine hit of completing a task and shipping something real to users. That kind of quick feedback loop is much harder to come by in leadership work, where progress is slower and impact unfolds over time.

Focus on the difference between the work you enjoy and the results that motivate you. Most meaningful work comes with some grind; the key is knowing which trade-offs you're OK with.

Ask yourself: "What outcomes feel most rewarding?," "What parts of the process am I willing to tolerate?," and "What stays energizing, even when it's hard?" If you had a day full of just one kind of task, which type would leave you feeling accomplished and motivated? Pay attention to those moments; they're pointing you toward the kind of work you may want more of in the future.

What felt most draining

Just as important is identifying what drained you. What parts of the role made you feel stuck, frustrated, or depleted?

For some tech leads, it's endless context switching. For others, it's people challenges that never seem to resolve. Maybe you found stakeholder alignment exhausting or struggled with the lack of deep technical focus. The point is to notice patterns.

Ask yourself: was it draining because I lacked the skills, or because the task itself didn't suit me? That distinction matters because if it's a skill gap, you can grow. But if it's a values mismatch or a persistent energy drain, it might signal the kind of work you don't want to prioritize going forward.

How the role has changed you

Being a tech lead changes how you think. It forces you to zoom out, make trade-offs, and weigh people and delivery with equal weight. You likely developed new muscles: influence without authority, coaching under pressure, managing complexity across people and systems.

How has your view of leadership evolved? How do you show up differently now than when you first started? What would your past self find surprising about how you operate today?

Signs of growth

It's easy to overlook your own growth, especially in a fast-paced environment. Take a moment to recognize how far you've come.

What felt hard or intimidating at first that now feels manageable or even easy? Maybe it was having tough conversations. Maybe it was leading architecture reviews. Maybe it was just knowing how to spend your time each day without a task list.

What feedback have you received, even informally, that points to your growth? Sometimes a team member saying "Thanks, I felt really supported" is more telling than a formal review.

Try this: write down three specific examples of things that used to feel difficult but no longer do. Include what changed, such as skills you built, confidence you gained, or support systems you leaned on. These stories are powerful evidence of growth, and they'll be helpful when building your development plan or preparing for performance reviews.

What you're curious about next

Finally, let your curiosity guide you. What do you want more of? Less of? Are there parts of this role that sparked interest in something else— product, engineering management, staff engineering?

You don't need all the answers right now. But you do need to start asking the right questions and paying attention to the patterns.

Some people prefer consistent tracking over a single big reflection moment. Daily journaling or regular check-ins can give you a clearer picture by capturing what's true in the moment, rather than relying on memory or standout experiences. It's a more accurate way to spot patterns, like what energizes you, what drains you, and how those shifts show up day-to-day, but it does require more consistent effort.

Whatever approach you choose, the goal is the same: to gather the raw material for better career decisions and to honor the journey you've already taken.

IDENTIFY YOUR STRENGTHS AND AREAS FOR IMPROVEMENT AS A LEADER

Once you've taken time to reflect on what you enjoy and don't enjoy about the tech lead role, the next step is to translate that reflection into clarity about your strengths and areas for growth.

The better you understand your current shape as a leader, the better equipped you'll be to make thoughtful career choices and design your growth path.

This clarity ensures that you choose a next role that plays to your strengths. It allows you to focus your learning and development time more effectively. It helps you have more informed and confident career conversations. And it gives you the foundation to advocate for yourself and track your own progress over time.

A simple way to start is by using tools like VIA Character Strengths (*https:// www.viacharacter.org*), CliftonStrengths/StrengthsFinder (*https://oreil.ly/4lVWa*), or a Work Styles Assessment (*https://oreil.ly/fcoQf*). While many people are understandably skeptical of these tools, as they're not scientifically rigorous, they can still offer real value when used thoughtfully. At the very least, they prompt useful self-reflection and can surface patterns or themes you might not have noticed on your own.

Just be mindful not to treat them as fixed truths or personality labels. These tests don't define you; the outcome is only as helpful as the insight it triggers. Focus on what resonates, discard what doesn't, and remember that context always matters: the same strength can show up very differently depending on your role, team, or environment.

Even disagreeing with the results can be eye-opening. It can help clarify how you see yourself or the kind of leader you want to become. These tools aren't definitive, but they can trigger meaningful conversations, especially with yourself.

Use a combination of self-reflection and external input to map your strengths and gaps. Building on the insights from reflective tools and feedback, this step allows you to connect what you're learning about yourself with how others experience your impact. It bridges internal awareness with external perception, giving you a more accurate, well-rounded picture of where you currently stand and where you can grow next.

What have you consistently done well? Think about the moments when others turned to you. What were you doing? Were you mentoring, solving hard tech problems, calming things during an incident, facilitating cross-functional planning? What do people often thank you for? What do team members and stakeholders rely on you for? What types of challenges do you handle with ease?

What feedback have you received? Review feedback from multiple sources, such as performance reviews, retrospectives, one-on-one check-ins, informal praise, or constructive notes from peers. Look for patterns. One comment might be an outlier, but repeated themes are signals. If you don't have feedback handy, ask. Reach out to a few trusted peers or collaborators and say, "I'm reflecting on my leadership journey. Is there anything you've noticed that I tend to do well or could do better?"

What challenges have you struggled with? What drained your energy, felt consistently difficult, or led to less-than-ideal outcomes? Sometimes these are skill gaps. Other times, they signal misalignment with your natural strengths or interests. What situations leave you second-guessing yourself? Where do you often need to ask for help? What have you avoided doing, even though you knew it mattered? This isn't a moment to be self-critical; it's a moment to be curious. Why do these things feel hard? Is it something you want to get better at? Or something you want to do less of in the future?

Whatever you discover in this reflection process, take a moment to celebrate. What's easier now than it was six months ago? What used to feel hard that now feels natural? Maybe you finally feel comfortable leading architecture reviews. Maybe you ran a retro that shifted how your team works. Maybe you gave feedback that landed well.

Those are signs of growth. Write them down. They're proof of your evolution, and they'll be useful when it's time to build your growth plan for your next move. (More on building a growth plan in the section "Developing a Personal Growth Plan" on page 38.)

Possible Career Paths After Tech Lead

Once you've taken the time to reflect on your experience as a tech lead—what energizes you, what challenges you, and where your strengths lie—the next natural question is: where do I go from here?

The good news is, there's no single path forward. The tech lead role sits at a unique crossroads between technical depth, people leadership, and organizational influence, and from here, your next step can take many forms.

Some tech leads transition into engineering manager roles and focus more on people and team development. Others lean into technical mastery by stepping into Staff+ or architect roles. Some shift toward coordination and execution as technical program managers, and a few even branch out into consulting or advisory work, bringing their experience to a broader set of teams.

In this section, I'll explore these options: what each path looks like, what types of people tend to thrive in them, and how to recognize which might be right for you.

ADVANCING TO ENGINEERING MANAGER

If you found yourself enjoying the people side of the role while reflecting on your experience as a tech lead, then a transition to the EM role might feel like a natural next step. In fact, it's one of the most common career moves after being a tech lead. The EM path is especially suited for those who are energized by growing people, shaping team culture, and influencing broader organizational direction.

Before I go into describing what the EM role is and how it's different from the tech lead role, it's important to mention: just as there is a lot of confusion around the tech lead role in the industry, the same is true, though perhaps less talked about, for the EM role. While EMs typically sit one level higher in the organizational hierarchy, the debates that surround the tech lead role also show up here: How technical should an EM be? How hands-on? How involved should they be in day-to-day team operations?

The answers vary widely. The definition of the EM role depends heavily on how many teams they manage, whether those teams already have tech leads, and the broader structure of the organization. When teams are missing strong tech leads, the EM often ends up pulled into more technical or operational detail than is ideal.

My personal take is similar to how I view the tech lead role: the EM role should be a highly people-focused one. But that view usually works better under a specific setup, one I've seen succeed repeatedly in my experience. In this setup, each team has a strong tech lead, and the EM works closely with them, focusing on the broader organizational view, people development, and long-term team health. Ideally, an EM supports multiple teams, allowing a clear separation between the responsibilities of EMs and tech leads.

In my experience, any extreme, whether it's having no tech leads at all, only EMs managing multiple teams, or having every tech lead report directly to the head of engineering, can quickly become overwhelming and unsustainable. I once consulted for a startup that had no tech leads, just EMs overseeing several teams. One EM was responsible for three different teams and was constantly pulled between technical decisions and people management. She was excellent at her job but visibly overworked. It was clear the absence of tech leads was creating unnecessary strain, both for her and for the teams that lacked day-to-day technical leadership.

Of course, there are many functioning org structures out there with different formats. That doesn't make one model better than another, just different.

It's also worth noting a current trend in the industry. At the time of writing, particularly with the AI boom, more and more EMs are expected to be close to the code and even to be contributing regularly. Personally, I disagree with this expectation. I don't believe EMs should be required to write production code. In fact, I think an EM coding outside of a team's regular process can do more harm than good. There are better ways for an EM to use their technical expertise: participating in high-level architectural decisions, helping with planning and expectation management, supporting tech leads, and ensuring long-term technical sustainability.

That said, I'm open to being proven wrong. Each company and team operates differently, and sometimes experimentation leads to surprising results. But as a default, I advocate for a hands-off EM model, one rooted in trust, strategic thinking, and team enablement.

So, what does the EM role actually look like in practice?

Based on the points I outlined earlier, the key differences between the tech lead and EM roles fall into a few major categories.

First, the level of influence. EMs typically sit one level higher in the organizational hierarchy, which gives them access to more upper leadership, broader visibility, and earlier awareness of company-wide changes. They're often included in planning conversations that shape future direction months in advance—things tech leads may learn about only once they begin impacting delivery.

In addition, the nature of stakeholder relationships shifts. While tech leads primarily collaborate with their immediate team and cross-functional partners like product, design, and their EM, engineering managers engage with a broader and more senior set of stakeholders, including HR, finance, other EMs, and senior leadership. The stakes are also higher: conversations often center on resourcing, organizational design, business risk, and long-term strategy. As a result, EMs must adopt an even more business-oriented communication style, framing decisions in terms of impact, return on investment, and alignment with company objectives.

Second, the scope of involvement. A tech lead should focus deeply on a single team, working closely with engineers in the day-to-day, guiding technical decisions, and unblocking delivery. An EM, by contrast, may support multiple teams. That broader scope means they're less embedded in any one team's daily challenges and more focused on patterns, people dynamics, and long-term team health across the board.

Third, decision-making power. While tech leads often provide input on things like performance reviews, compensation, and growth opportunities, EMs usually make the final calls. They're the ones responsible for promotions, salary changes, budget allocation, and sometimes having tough conversations around underperformance. A strong tech lead will guide and support their team members closely, but it's often the EM who formalizes and delivers the outcomes.

Finally, there's a difference in relationship dynamics. Tech leads are in frequent one-on-ones with their team members; those conversations are core to the role. With EMs, one-on-ones with individual engineers might be less frequent, especially when teams have active tech leads. Instead, EMs are likely to have more regular syncs with tech leads themselves, supporting them in supporting the team.

So the real question is: are you ready to move even further away from the code?

If yes, the EM path might be a great fit. But if you feel more fulfilled working through complex technical challenges yourself and staying closer to the code, you may want to explore staff engineer or architect paths instead, which I'll cover next.

SPECIALIZING AS A STAFF+ OR ARCHITECT

If the EM role doesn't sound like a fit—maybe you're not drawn to managing people's development, handling performance reviews, or distributing budgets—and you'd prefer to remain on the technical track, then a Staff+ role, such as staff engineer or architect, might be a better path moving forward. These roles allow you to stay close to the technology, focus on complex engineering challenges (driving major infrastructure migrations, solving critical reliability issues), and lead through technical expertise rather than through direct management.

Note that when I refer to Staff+, I'm talking about the family of senior individual contributor roles that typically include titles like staff engineer, senior staff, principal engineer, and distinguished engineer. While the naming and levels vary across companies, what they share is a common idea: as you progress through these levels, your influence grows from the team level, to multiple teams, and eventually to the organization as a whole. With each step, the expectation is that your impact increases accordingly—not just through individual contributions but through broader technical leadership.

Just like with the tech lead role, there are varying opinions and definitions out there for what these roles entail. To explain the range of perspectives, I'll reference a few voices that have articulated them clearly.

There are a few defining traits that the Staff+ and architect roles have in common, starting with "You're not a manager, but you are a leader." This is one of the key takeaways from Tanya Reilly's *The Staff Engineer's Path*, and it applies just as much to architects as it does to staff engineers. Even without direct reports or formal managerial authority, these roles come with expectations of high impact, maturity, and influence.

Reilly puts it clearly: "Staff engineers lead differently than managers.... Their impact happens in other ways." Staff engineers are expected to be the grown-ups in the room: calm in a crisis, grounded in judgment, and trusted to lead without needing authority. Their leadership shows up not in performance reviews or approving time off but in how they shape systems, mentor team members, make technical calls, and guide the team through complexity.

At this level, technical expertise is just the foundation. You're also expected to influence cross-team decisions, align with business needs, and communicate clearly with diverse audiences. Besides deep technical skills, these roles demand collaboration, teaching, and strategic thinking.

This aligns with the view of leadership described by Will Larson in his article "Staff Archetypes" (*https://oreil.ly/K-ksY*): leadership can look like mentoring, shaping patterns, influencing without authority, and raising the quality bar across the board.

The main differences between staff engineer and architect roles tend to center on scope, visibility, and approach.

Staff engineers are typically embedded within one or a small number of teams, where they focus on high-impact technical work. They're deeply involved in day-to-day engineering challenges, reviewing designs, writing code, mentoring peers, and unblocking complexity as it arises. They're often seen as domain owners: the go-to experts for a specific technical area.

In smaller or less mature organizations, these responsibilities can overlap with those of a senior engineer. The key difference often comes down to scope, influence, and consistency of impact. While a senior engineer might lead a project within the team's scope, a staff engineer is expected to lead across boundaries, influencing technical direction beyond their immediate team and operating with a broader, more strategic lens.

Architects, on the other hand, tend to operate across multiple teams or even entire departments. They define system boundaries, maintain architectural consistency, and guide long-term evolution across projects. They focus more on big-picture design, scalability, and aligning technology strategy with business goals.

Will Larson describes the architect role as an archetype of the staff engineer role. In his view, architects are responsible for the success of a particular technical domain, such as cloud infrastructure or API design, and that domain must be complex and central enough to justify this kind of specialized ownership. He also counters the negative stereotype of architects working in isolation. Effective architects, he says, are embedded in the business context and earn influence through judgment and collaboration.

Another key difference is how hands-on the role remains. Staff engineers are expected to be close to the codebase, while architect expectations vary by organization. Some companies expect architects to write code regularly; others want them focused on design and alignment. Both models can work; what matters is clarity on the expectations and a good match between the role and your strengths.

So how are these roles different from the tech lead role? Think of the tech lead as sitting somewhere between the people-focused EM and the technically deep staff engineer or architect. Tech leads are often responsible for delivery within a single team, balancing both technical guidance and day-to-day coordination. In contrast, Staff+ and architect roles lean much more deeply into the technical end of the spectrum. They're expected to drive technical direction across teams, lead through influence, and solve large-scale engineering problems without managing people directly.

So if you want to go even deeper into technology, and you enjoy leading through influence, design, and problem-solving more than through people management, Staff+ or architect paths might be a strong fit. For many, it's an ideal way to stay rooted in technical work while continuing to grow their impact across the organization.

TRANSITIONING TO A TECHNICAL PROGRAM MANAGER ROLE

If, as a tech lead, you found yourself energized by driving alignment across teams, navigating complex project scopes, or orchestrating delivery plans but felt less attached to deep technical ownership or hands-on coding, the technical program manager (TPM) path might be a natural evolution.

TPMs are responsible for managing large-scale, cross-functional initiatives that require coordination across engineering, product, design, and business teams. Their superpower is anticipating risk, aligning stakeholders, managing dependencies, and keeping delivery on track, especially in fast-moving or high-complexity environments.

While they're not expected to code, TPMs typically have strong technical backgrounds. This technical fluency helps them understand trade-offs, ask the right questions, and connect the dots between engineering decisions and business goals. In many ways, a former tech lead stepping into a TPM role already has a head start: you're used to operating at the intersection of technology, product, and execution.

TPMs shine in situations where the scope spans multiple teams or systems, where aligning various stakeholders on timing and outcomes is critical, where coordination costs are high or timelines are long, and where product and engineering teams need structured delivery support.

One important distinction: while TPMs often collaborate closely with engineering managers and tech leads, their focus is more on program execution than people development or system architecture. That said, strong TPMs bring leadership, clarity, and calm to situations where ambiguity is high and stakes are even higher.

If you enjoy steering planning sessions, solving organizational blockers, and being the glue across teams but don't feel the pull to manage people or deepen your technical specialization, the TPM role can offer high impact and visibility, without requiring a move into traditional management or staying deep in the code.

GETTING INTO CONSULTING OR ADVISORY ROLES

If you've developed strong expertise as a tech lead and enjoy solving complex problems across a variety of contexts, a move into consulting or advisory work could be a compelling next step. These roles let you apply your knowledge more broadly, often across multiple teams, organizations, or industries, without being tied to a single delivery team or company structure.

Consultants are typically brought in to provide clarity, solve thorny problems, or accelerate transformation. You might be asked to evaluate architecture, improve delivery practices, or help a struggling team level up its ways of working. Advisory roles are similar but often longer-term and more strategic, and sometimes at the executive level, helping organizations make high-impact technical or organizational decisions.

This path is a great fit if you enjoy stepping into messy situations, quickly making sense of unfamiliar systems, and having a bird's-eye view of how different parts of a business connect. It's also a good option if you want more flexibility or autonomy in your work, as many consultants and advisors work independently or in smaller specialized firms.

But it's not for everyone. Consulting requires high adaptability, strong communication skills, and a certain level of emotional intelligence. You'll often be an outsider, so building trust quickly is critical. You also need to be comfortable with ambiguity and able to deliver value without always having full control over execution.

Still, for experienced tech leads who want variety, influence, and the chance to work across broader scopes, consulting or advisory roles can offer rich and meaningful ways to grow their careers.

Take my own journey as an example. After working as a tech lead at Thoughtworks, I transitioned into being a full-time consultant, trainer, and career coach. I realized that the skills I had built over the years, from mentoring and stakeholder management to systems thinking and communication, could bring value in different forms beyond a single team or company.

Over the past four years, I've coached more than 500 professionals across levels and disciplines, trained over 300 tech leads on building high-performing teams, and worked with a number of companies to improve engineering practices and leadership culture. I've also built a community of nearly 30,000 people by sharing stories and lessons learned on LinkedIn and through my newsletter.

I share this not to position my path as the path but to highlight that the tech lead role often prepares you for more than you realize. It can be a powerful springboard into a wide variety of roles or even portfolio careers that align better with your goals and values over time.

The key takeaway: the skills you're developing now—solving problems, growing people, aligning stakeholders, communicating clearly—are valuable well beyond your current title. You may be closer than you think to your next career move.

Planning Your Transition

Once you've identified a direction that feels aligned with your strengths and interests, the next step is turning that vision into action. Career growth doesn't happen all at once; it happens through intentional choices, small experiments, and clear preparation.

Whether you're aiming to become an engineering manager, a Staff+ engineer, a TPM, or something else entirely, the way you approach the transition matters. This section is about helping you bridge the gap between where you are and where you want to be.

I'll talk about how to choose your next step with intention, how to prepare yourself for the shift (mentally, emotionally, and practically), and why it's OK if your path isn't linear. Most importantly, I'll remind you that you're not locked in, and you have more flexibility than you think.

CHOOSE YOUR NEXT STEP WITH INTENTION

Now that you've reflected on what energizes you, where you thrive, and the kind of impact you want to have, it's time to make a conscious choice. Whether you feel drawn to people leadership, technical depth, strategic coordination, or advisory work, what matters most is that your next step aligns with your values and aspirations.

You don't need to have everything figured out. What's important is that you're moving forward with self-awareness and curiosity, not just following the default path. Your time as a tech lead has given you insight into what you enjoy, where you make the biggest impact, and where you want to grow. Let that learning guide your next move.

That reflection sets the foundation. Now it's time to prepare for the shift.

PREPARE FOR THE SHIFT

No transition happens overnight. Once you've gained clarity about the role you want next, the real work begins: identifying the skills, experiences, or relationships you need to develop and starting to take intentional steps toward them.

Sometimes, that means changing what you prioritize. I was once coaching a tech lead who wanted to become an engineering manager. She had a clear long-term goal, but her day-to-day choices didn't reflect it. She was deeply focused on her own team, spending most of her time coding and supporting them directly, and rarely engaging beyond that. In her words, she would "default to contributing to a feature over joining a cross-team conversation."

When she realized that her current behavior wasn't aligned with her future aspirations, she shifted her approach. She made a plan: reach out to other engineering managers, build relationships, and get involved in cross-team initiatives. This meant stepping back from some coding and resisting the urge to jump into execution mode every time the team hit a bump.

It was hard at first. She struggled with guilt about "not being helpful" or "not contributing enough." But the more we revisited her goal, reviewed how she spent her time, and talked through why this shift mattered, the easier it became to stay on track. Over time, her habits aligned more closely with her ambitions, and she started to see real progress.

If you're preparing for a transition, here are a few ways to start:

Start by auditing your own calendar

Inspired by the preceding story, take a look at how you've spent your time over the last one or two weeks. What types of activities are taking up most of your energy? Are they aligned with the role you're aiming for?

This kind of audit can help you spot where your time and attention are going and whether your current behavior supports your long-term goals. If not, it's a good starting point for change.

Make a growth plan toward the role you want

Start writing down the steps you need to take to reach your next role. Define the skills that qualify someone for that position, and reflect on which of those you already have and which you might need to develop. Identify people around you who can help you get there and start building those relationships. The section "Developing a Personal Growth Plan" on page 38 covers more on how to do this effectively for the tech lead role, but the principles can be applied to any role.

Start having conversations

One simple thing that's often overlooked: if you have clarity about where you want to go, don't keep it to yourself. Tell the people who can support or influence your next step.

Start with your manager. Many people feel like it's not OK to speak up, but great managers are usually relieved and grateful when you do. It makes their job easier: no need to dig, guess, or push. They can focus on helping you grow in the direction you actually want.

Talk to people who are already in the role you're aiming for. Ask about their day-to-day work, their biggest challenges, and the skills they rely on most. Try questions like "What surprised you most about the role?," "What do you wish you'd known before stepping into it?," "What should someone focus on when preparing for this kind of work?" Pay attention to the patterns that emerge, especially around what strengths you already have and where you might want to grow.

And finally: tell anyone who's asking. If someone opens the door, even casually, take the opportunity to share what you're aiming for. You never know who might help open the next one.

Shadow or partner

Find opportunities to work with people in the role you're aiming for. This can mean joining a project they're leading, sitting in on their planning sessions, or even just grabbing coffee to hear about their experiences. If possible, try to build a relationship with a mentor who already holds the role you're interested in. Having someone inside the company who understands the expectations, challenges, and opportunities can be incredibly valuable. They can help you gain visibility, recommend you for stretch opportunities, and offer feedback that accelerates your growth. Don't just look for someone to share tips; seek out someone who can champion your growth and help create real opportunities for you.

Say yes to an initiative that aligns with your next step

This could mean leading a cross-functional project, mentoring someone outside your team, or taking the lead in planning discussions that span teams.

If you want to move toward a more technical role, look for opportunities where you can demonstrate architectural thinking or contribute to system design. If you're more interested in a people leadership path like engineering management, seek coordination roles, mentor more junior engineers, or cofacilitate team rituals.

And if you're intrigued by the TPM path, use your tech lead day-to-day responsibilities as a springboard, and maybe work more closely with your project manager. Take on more ownership in planning work, offer to manage timelines or dependencies across teams, and step into conversations that require navigating ambiguity.

Look around your current responsibilities. Chances are, there are already ways to explore these shifts within your existing role.

These are all ways to stretch into the new role before formally stepping into it.

Expand your horizon

Get outside of your current environment and company. Learn what other companies are doing, and explore different definitions and formats of the roles you're interested in. Keep in mind that there are opportunities to grow beyond your current company. I know tech leads who were told they'd have to wait years for a transition internally but moved into new roles almost immediately elsewhere. And as I've kept repeating throughout this book, roles mean different things in different companies, so find the environment that defines the role in a way that matches what you actually want to do.

Give yourself a confidence boost

Shifting toward a new role can trigger uncertainty, especially when you're stepping into something less familiar. It's normal to doubt yourself or feel like you don't measure up yet. Confidence, though, isn't something you either have or don't; it's something you can actively build.

When I work with tech leads in transition, I often recommend building small, repeatable habits that reinforce your strengths. That could be reviewing positive feedback you've received, keeping a running "done list" to remind yourself of what you've already achieved, or writing down recent wins—even small ones—at the end of the week. Some people record voice notes to talk themselves through tough days; others visualize how they'd show up in the new role and use that to shape their actions now. The key is not to wait for confidence to appear but to practice the behaviors that build it.

If you want more concrete strategies to try, I've written a guide (*https:// oreil.ly/Vwve6*) that collects several of these.

This transition is an evolving process. Keep checking in with yourself, stay flexible, and remember that career paths are built step-by-step.

REMEMBER: YOU'RE NOT LOCKED IN

People are often afraid of making the wrong move. But when you don't know exactly what's right for you, any choice that moves you forward is the right one.

For example, some tech leads decide to go back to being senior developers after trying the role. They often ask me: "Is this a step back in my career?" My answer is always the same: there are no steps back from here. Careers aren't linear; they're made up of experiences that help you learn what fits and what doesn't. Even trying something and realizing it's not for you is a step forward in understanding what you truly want.

It's OK if your next step doesn't go exactly as planned. What matters is that you keep learning, stay honest with yourself, and remain open to new possibilities. No single role defines your career. Each one is just a chapter in a much longer story.

And remember, no decision is forever. We are lucky that in the tech industry, we can more easily move back and forth between roles. There's room to explore, adapt, and reframe our path based on evolving interests and needs. I know tech leads who returned to full-time development because they preferred it. I also know others who moved into EM, TPM, or even product roles with ease.

The most powerful thing you can do for your career is keep moving forward with intention, curiosity, and self-awareness.

Key Takeaway

If there's one thing to take away from this chapter, it's that there is no single "right" next step after being a tech lead, only the one that's right for you.

Whether you choose to stay in the middle—continuing to grow in the tech lead role—lean further into people leadership, deepen your technical influence, or explore something entirely different, the reflections you've done in this chapter are your best guide forward.

The real win is choosing with intention: based on what energizes you, where you want to grow, and the kind of impact you want to make.

Careers are long. Paths are flexible. And you're allowed to evolve. So wherever you head next, stay curious, stay grounded in what matters to you, and trust that every step forward is part of building a career that truly fits.

Wrap-Up

If you've made it this far, thank you. Writing this book was my way of distilling years of questions, hard lessons, and cherished victories, and my hope is that it helped you feel a little less lost in your tech leadership journey.

We've covered a lot, from the messy middle of transitioning into the tech lead role, to managing people, projects, and stakeholders, to figuring out what comes next. But if there's one message I want to leave you with, it's this:

Leadership isn't a destination. It's a series of decisions.

Decisions about how you show up for your team. How you grow. Where you invest your energy. And eventually, what you want your career to look like beyond the role you're in today.

Your path won't look like anyone else's. And it shouldn't.

There is beauty in ambiguity and complexity. When the role isn't clearly defined, it means you have a chance to shape it, to make it your own. That freedom can feel daunting, but it also means you get to build something that fits you and serves your team in the best way possible.

This may be the end of this book, but it's not the end of your journey. In fact, it might just be the beginning.

I continue learning every day about this role, through experience, conversations, and every tech lead I meet. I'll keep sharing what I learn through my newsletter (*https://oreil.ly/ckH8U*). Follow along if you'd like more templates, tools, stories, and reflections on these topics.

And if you ever have a question, want to share your perspective, or just want to connect, don't hesitate to reach out (*https://oreil.ly/GNqCi*). I'm always open to new ways of thinking—and who knows, maybe you'll even change my mind about something. ;)

Good luck!

Index

About the Author

Anemari Fiser is a tech leadership trainer and coach who helps engineers grow into confident and effective tech leads. With over a decade of experience in the tech industry, she has held roles ranging from software engineer and consultant to tech lead and engineering leader before dedicating her career to developing others.

Over the past four years, as an independent professional, Anemari has coached more than 500 engineers and partnered with companies to train nearly 400 tech leads on building high-performing teams, improving communication, and leading with confidence.

She is the creator of O'Reilly's *Soft Skills for Tech Leads* course and the *Level Up as a Tech Lead* newsletter (*https://level-up-as-a-tech-lead.anemarifiser.com*), read by 7,000+ tech professionals worldwide. Anemari also shares practical leadership insights with a community of over 30,000 engineers on LinkedIn, where she writes about the real challenges of leading in tech.

Her practical, people-centered approach helps technical leaders navigate the challenges of leadership while keeping their authenticity—and their sanity—intact.

Colophon

The cover illustration is by Susan Thompson. The cover fonts are Gilroy Semibold and Guardian Sans. The text fonts are Adobe Myriad Pro, Adobe Minion Pro, and Scala Pro, and the heading font is Benton Sans.

O'REILLY®

Learn from experts.
Become one yourself.

60,000+ titles | Live events with experts
Role-based courses | Interactive learning
Certification preparation

Try the O'Reilly learning platform free for 10 days.

www.ingramcontent.com/pod-product-compliance
Lightning Source LLC
Chambersburg PA
CBHW060755220326
41598CB00022B/2439